THE AVRO
SHACKLETON
GUARDIAN OF THE SEA LANES

A stunning photograph from the skilled hands of
Charles E Brown showing the prototype MR Mk 3, WR970 in flight.
This is a very early image, as the stall warning strips later fitted
to the inboard wing leading edge are not present.

RICHARD A FRANKS

DALRYMPLE
& VERDUN◆
PUBLISHING

Command of the Sea

NEW ZEALAND

Maritime Reconnaissance Bomber is the title for an aircraft which must be capable of fulfilling a variety of roles. That is why Coastal Command Royal Air Force, is now being re-equipped with Avro Shackletons, specially designed and fully equipped for tasks as arduous as any in the history of flying. The provision of four engines, regarded as an essential, inspires confidence in the crew, whose comfort and requirements have been main considerations in the layout of its spacious interior. With the most up-to-date radio and radar, with its powerful offensive armament and low level manoeuvrability coupled with its ability to carry large and varied loads to extreme ranges, the Avro Shackleton is a formidable attacker.

Shackleton

A V ROE & CO LTD MANCHESTER. *Member of the Hawker Siddeley Group*

CONTENTS

The Avro Shackleton
Guardian of the Sea Lanes
Richard A Franks

ISBN 1-905414-01-3

First published in 2005 by
Dalrymple & Verdun Publishing
33 Adelaide Street, Stamford
Lincolnshire PE9 2EN
Tel: 0845 838 1940
Email: info@dvpublishing.co.uk
Web: www.dvpublishing.co.uk
© Concept and design
Dalrymple & Verdun Publishing and
Stephen Thompson Associates.
© Richard A Franks 2005
© Richard J Caruana - Colour Profiles

© Crown Copyright used with the
permission of Her Majesty's Stationery
Office (HMSO)

The right of Richard A Franks to be
identified as the author of this work has
been asserted in accordance with
sections 77 and 78 of the Copyright
Designs and Patents Act, 1988.

Printed in England by
Ian Allan Printing Ltd
Riverdene Business Park, Molesey Road
Hersham, Surrey, KT12 4RG

Acknowledgments
I would like to express my thanks to Tiny
Potter, editor of The Growler produced by
the Shackleton Association, for all his
help and assistance in tracking down
information and people, to David Leppard
for taking the time while on holiday to
write the introduction for this title, to
Richard J Caruana for all his help, support
and those stunning colour profiles and
Godfrey Mangion, Christopher Ewing,
Philip Moody, Glen Cameron Williger, John
Woodward, Steve Vessey, Andrew White.
Keith Cheshire, Laurie Bruce, Trevor
Dobson, Barry Jones and Martyn Chorlton
for helping me with photographs.

For anyone with an interest in the 'Old
Grey Lady' we would recommend that you
consider joining the Shackleton
Association. This is a charitable
organisation run by a dedicated band of
ex-Shackleton crews as well as those who
built and serviced the type. For more
details visit their website at:
www.shackletonassociation.org.uk,
where there is a link to the membership
secretary and details of how to join.

PREFACE

I first met the 'Old Grey Lady' at RAF Ballykelly in 1961 and my first impressions were hardly favourable. At first I found her difficult to fly: the rudders and elevators were effective, but the ailerons were not. The Griffon engines were powerful, but not always reliable: I suffered no less than 28 engine failures during my career and two engine fires. However, getting home on three was no real drama. My affection for the aircraft grew with time, and now I have nothing but praise for her. The course at RAF Kinloss taught me not only to fly the aircraft competently, but also the art of working as part of a crew of 10. Any Shackleton crew member will tell you that the whole essence of being in the 'Kipper Flee', as we were then known, was that you worked as a unit: at the end of the day your life depended on the skill of all those with whom you flew. Our bread and butter in the 1960s was anti-submarine warfare and maritime surveillance, but because of the Shackleton's uniqueness: long range, large carrying capacity, and ability to deploy with our own ground crew, we undertook many varied tasks. To name a few: hurricane relief in Belize; support of British nuclear tests; support of the navy during The Indonesian Confrontation operating out of Borneo, medium level bombing with 1,000lb bombs in The Radfan; support of The Rhodesian oil embargo operating out of Madagascar; and following the conversion of role to Airborne Early Warning, air defence. Finally, there was Search and Rescue. I was captain of the first Shackleton to respond to the liner Lakonia on 23rd December 1963. Although fire totally destroyed the ship and many lost their lives the pure satisfaction of contributing to the rescue of hundreds made all the boredom of staring at an empty sea for hours on end, and long detachments from families to far flung locations, worth while.

David Leppard

Above: A view of the passenger ship Lakonia, which caught fire on the 23rd December 1963, is seen here in a photograph taken by David Leppard, the pilot of Shackleton WL757, which was the first aircraft on the scene.

Opposite page top: MR Mk 3 prototype WR970. The various changes made to create the Mk 3 are clearly shown here, although WR970 never received any of the Phase modifications of the rest of the fleet, as it was tragically lost during initial acceptance trials.

Opposite page bottom: MR Mk 1 prototype VW131 and going by the mix of other types in the background coupled with the barriers and tents my guess is that this photograph was taken at the SBAC event at Farnborough. *Both via Martyn Chorlton*

David Leppard operated the Shackleton for many years and accumulated some 5,000 hours on the type, all with the Mk 2 or AEW Mk 2. His service association with the aircraft was as follows.

1961 No.210 Squadron, RAF Ballykelly.
1962 MOTU Course, RAF Kinloss.
1962-65 No.224 Squadron, RAF North Front, Gibraltar (Captain during 1963-65 period).
1965-67 No.204 Squadron, RAF Ballykelly, (captain of the first constituted crew).
1970-71 OC No.204 Squadron, RAF Honington.
1971-72 Flight Commander Operations, No.8 Squadron, RAF Kinloss/RAF Lossiemouth.
1988 Last trip in a Shackleton as Commandant CFS checking squadron QFI.

Chapter 1: **THE BIRTH OF COASTAL COMMAND**

With the cessation of hostilities in 1918 and the formation of the RAF the period that followed saw the development and steady increase of the service. This increase did not however envisage another war and as such was very much based on an increased home defence force coupled with forces to meet colonial 'policing' demands elsewhere. The UK based element was known as the Home Defence Force and in 1925 this was placed under a new single command structure that was called the Air Defence of Great Britain. Within the command were the Auxiliary Air Force and Special Reserve units, but in July 1936 Sir Hugh M Trenchard, Commander in Chief of the RAF, decided to dissolve this command and reform it as Fighter Command. This new command included the fighter elements combined with the Army Co-operation units and the Royal Observer Corps, Coastal Com-

March 1939 they had converted to the Supermarine Stranraer. The type more commonly associated with Coastal Command is the Sunderland and this came from Specification No.R.2/33, which called for an anti-submarine and ocean patrol flying boat based on the successful Short S.23 commercial 'C' Class flying boat. The S.25 resulted from the specification and was adopted for service use as the Sunderland Mk I; the first prototype flying on the 16th October 1937. No.210 Squadron was the first to receive the type in May 1938, followed by No.230 at Seletar in June and in 1939 Nos 204 and 228 also received the type. By the time hostilities flared up again in September 1939 Coastal Command had all of the above squadrons plus No.22 at Thorney Island and No.42 at Bircham Newton that were designated as 'strike' units, even though the type that they operated in this offensive role was

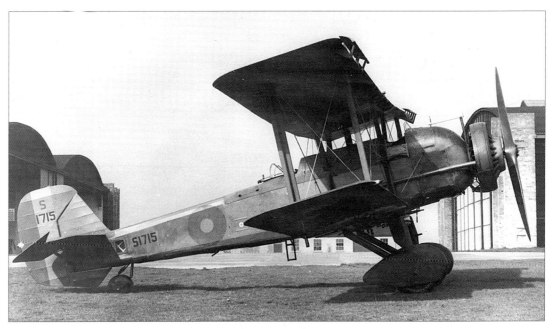

Vickers Vildebeest Mk I
S1715. *Author's Collection*

mand, Bomber Command, Training Command and the Fleet Air Arm. The latter caused considerable concern and a lot of lobbying by the Admiralty, which saw all operations in the vicinity of the sea as being their responsibility. As a result of their lobbying, on the 30th July 1937 the Fleet Air Arm was placed under the direct command of the Admiralty. Control of Coastal Command was also sought by the Admiralty, although this would not become apparent until later.

Coastal Command from the outset operated flying boats, and initially they had such types as the Saro London, which equipped Nos.201, 202, 204, 209, 210, 228 and 240 Squadrons. No.209 Squadron operated the Short Singapore initially, although by

the obsolete Vickers Vildebeest biplane! The command also operated the Avro Anson in the patrol role, as well as just starting to equip with the new Lockheed Type 14, which had been christened the Hudson.

It was obvious as soon as hostilities resumed to those in the Admiralty that once again Britain would be dependant on the supply of raw materials by sea. Their experience in the first few years of the First World War lead them to believe that once again a system of regular convoys would see the country through, and so very quickly this principle was put into action. It soon became apparent though that the range of the existing aircraft in Coastal Command was insufficient to meet the requirements of the new convoy system, as

New York was some 3,000 miles away from the nearest British port and the Sunderland only had an operational radius of 800 miles. German U-boats operating from the captured bases in France were soon taking a terrible toll of Allied shipping and so some form of new long-range anti-submarine aircraft was desperately needed. Initially it had been hoped that some of the new four-engine bombers being produced for Bomber Command could be spared, but this was not to be, as Bomber Command needed every aircraft that was made. Back in July 1939, the Air Ministry had approached the American Consolidated Vultee Aircraft Corporation to obtain an example of the Catalina flying boat for evaluation. Evaluation of this aircraft by the Marine Aircraft Experimental Establishment at Felixstowe lead to an order for fifty airframes. These started to arrive in 1941, with No.209 Squadron based in Northern Ireland getting the first of them. The other product that the Air Ministry procured from Consolidated was their Model 32 bomber. This type was initially ordered by the British as the Liberator, although later the USAAF also adopted the type as the B-24. The initial batch of twenty-six Liberators were envisaged for use with Bomber Command, but when they arrived in March 1941 they were soon found to be unsuitable for their intended role and were soon relegated to the transport role with BOAC. A revised version was ordered for Coastal Command, where it became the Liberator Mk I and was operated first by No.120 Squadron in Northern Ireland.

The Hunted becomes the Hunter

Having been at a disadvantage for the initial stages of the war, with the arrival of a long-range, land based anti-submarine aircraft, Coastal Command was at last able to go on the offensive. There was still a problem to overcome though, and that was actually finding a U-boat etc. in the massive expanses of the Atlantic. The use of radar is well documented elsewhere, but by August 1937 an Anson was fitted with a basic aerial radar system and was test flown by the A&AEE. Full development of this system was to prove a lot more troublesome than had initially been expected and it was not until 1939 that ASV (Air to Surface Vessel) Mk I radar was used in a Hudson to locate a vessel at sea (actually the submarine L27, which was detected while on the surface). With even greater aerial arrays fitted to a Whitley the effective range of the system was increased to 15 miles, although the RAE had developed an even better (1.7cm wavelength) system that had a range of 36 miles and this became ASV Mk II in early 1940. The Sunderland, Hudson, Wellington, Whitley and Liberator all had ASV Mk II fitted and it had its first operational success in a Whitley Mk V on the 30th November 1941 when it was used to find U-boat U-206, which the aircraft subsequently sank. Further development of radar and the introduction of the cavity magnetron oscillator lead to an ever increasing potential in radar detection. Coastal Command often

played a poor relation to Bomber Command though, with the latter getting all the new radar systems either for its bombers or for Fighter Command's night fighters. ASV Mk III was actually a less potent version of H2S and this type used the now-familiar 'Yagi' antennae (this being named after the Japanese electrical engineer Hidetsugu Yagi). The scanner for this system, with a reduced scanning radius of 60°, was installed under the nose of Wellington Mk XIIs and XIVs. An improved version, ASV Mk VI was soon produced and this was further improved with the Mk VIA that actually directed the Leigh Light at the target.

The Leigh Light was the brainchild of Sqn Ldr 'Sammy' Leigh. He took the 22.24in carbon-arc searchlight and developed it for airborne use. Initially the light was to have been installed in a retractable ventral machine-gun station, but this was never used operationally. The light was fixed in a small housing that was usually fitted under the starboard wing and the first operational success of the Leigh Light and ASV radar was on the 6th July 1942, when a Wellington Mk VII of No.172 Squadron located U-502.

Saving Lives

Most accounts dealing with Coastal Command operations in WWII tend to look at the offensive elements, however the command also undertook the vital role of Air-Sea Rescue. The airborne lifeboat was initially carried by the Hudson, but it was to be the 'replacement' for the Wellington, in the form of the Warwick, that most actively filled the role. The type had initially been refused by Bomber Command and it was not until 1943 that the Air Staff decided that the type would be ideal in the ASR role. Trials with an airborne lifeboat resulted in the type being put into production as the Warwick ASR Mk I, of which 204 were built. Later another 165 Warwick ASR versions were built, all with improved equipment and/or lifeboats

A WWII Perspective

WWII had highlighted the limitations of using flying boats for long-range reconnaissance in a war theatre

ASV radar antenna under the starboard wing of a Wellington Mk VIII.

that was ever-expanding. German U-boats had quickly operated deeper and deeper into the Atlantic ocean, and thus moved far beyond the effective operational radius of flying boats. The hasty conversion of four-engine bombers for the task of long-range reconnaissance and anti-submarine duties was initially, in the overall scheme, ineffective. Trying to use surface vessels to hunt and kill submarines was even less effective due to the fact that these ships, as they operate in the same area as the submarines they are hunting, run the risk of becoming the hunted. They are also slow and can only cover a small area effectively. The systematic development of better aircraft and the adoption of radar systems and developments like the Leigh Light all lead to a steady increase in the effectiveness of Coastal Command and by the wars end they had accounted for 196 German submarines, a figure that represented a major portion of the total number destroyed. Wartime experience had shown however that the use of land-based multi-engine bombers in the role of submarine hunter/killer was very effective. In comparison with a flying boat, a land-based aircraft could translate the large hull of the former into additional space for fuel and thus extend their operational range while still using the same engine power. Experience had certainly shifted the perspective with regard to long-range reconnaissance over water and the whole subject of effective anti-submarine warfare. All of the lessons learned were to set the scene for the development of a new generation of maritime aircraft, although in the immediate post-war period it was going to be a long and painful process.

A front view of an unidentified Wellington Mk VIII fitted with radar.

Liberator GR Mk V, the British equivalent of the B-24D, 112 were supplied to the RAF under the lend-lease scheme.
via Martyn Chorlton

The Lockheed Hudson, which originated from the Type 14.
via Martyn Chorlton

The Consolidated PBY Catalina in RAF markings.

Originally built as a B Mk 1, RE164 was converted to a ASR Mk 3 and is depicted here in its final role as a GR Mk 3. *via Martyn Chorlton*

A clear underside view of a Warwick Mk V in flight. Note the chin radome very similar to that which would be fitted to the initial Shackletons. *Author's Collection*

Chapter 2: **A LEGEND FROM A LEGEND**

With the end of World War II, all lend-lease aircraft were returned to America and many squadrons were quickly run-down as men returned to 'civvy street'. Initially it was projected that a maritime reconnaissance version of the Lincoln B Mk 3 would fill the vacated Maritime Reconnaissance role. Late production Lancaster B Mk III bombers had been converted by Cunliffe-Owen Aircraft Co. Ltd to the ASR role, and thus became ASR Mk IIIs. Many of these were further modified in 1947 to fulfil the General Reconnaissance (GR) role and were thus redesignated as GR Mk 3 (the official numbering system having gone from Greek [e.g. Mk III] to Arabic [e,g, Mk 3] during 1947). Considering the suitability, albeit in a modified form, of the Lancaster and the projected use of the Lincoln in the Maritime Reconnaissance role, it is no surprise that Avro received an almost 'off-the-drawing-board' order on the 21st March 1946 to meet Specification 42/46 for a new aircraft for this role. The initial order was for twenty-nine airframes, allocated serials VP254 to 268 and VP281 to 294 and there was also a need for three prototype airframes (VW126, VW131 and VW135). Procurement of new aircraft in the immediate post-war era was a slow process and a new system had been created to filter out unnecessary duplication of designs. This system was known as Operational Requirements (ORs) and consisted of a series of consecutive numbers starting at 100. Each OR could lead to further discussions, development or the requirement could be superseded or lapse. It was OR.320 that covered the need for a long-range maritime reconnaissance aircraft and it was this OR that Avro submitted their initial proposals for. Unfortunately it soon became apparent from Avro's initial design studies that a conversion based on the Lincoln would not work and the resultant aircraft would not be able to fulfil the requirements as identified in OR.320 satisfactorily. The design team at Avro actually went about designing a complete new aircraft to fill the revised (or should we say 'clearly defined') role and as a result the Air Ministry issued a revised specification number R5/46 to cover it. The term 'Lincoln 3' was used for quite a while to describe the Avro Type 696 that resulted from R5/46, however this was just a very loose term that belied the fact that the 696 was a completely new design. In designing their new aircraft Avro used a tried and tested formula that saw a large, lightly-loaded wing (from the Lincoln) with the revised undercarriage and elongated inner nacelles of the new Tudor airliner attached to a new load-bearing centre section. This new fuselage featured an entirely new front section forward of the wing plus a massive bomb bay. Its capacious interior easily housed all ten crew members, plus all the necessary sensors and weapons. Don Andrews the project designer at the time said that the revised fuselage was a simple conversion of the existing Lancaster/Lincoln one and that they just 'split it down the middle and made it about two foot fatter all round'. This is probably an oversimplification of the work involved, but the lineage of the type was plain for all to see. The design used several elements of the Tudor airliner, including the undercarriage/nacelles already mentioned, and this was mated to a high-set tail with large, oval-shaped end plates. Avro described the new design as consisting of 'Tudor outer wings and undercarriage, a modified Lincoln centre section, a wider and deeper fuselage and finally a Lincoln tail unit'. Again this is an oversimplified description, as certain areas like the tail were not directly taken from the Lincoln, as the end plates are much broader, rounder and do not have the distinct squared-off edge to the base of the trim tab seen in the later Lancaster marks

The second prototype, VW131, seen taking off from Farnborough during the SBAC show. *A V Roe*

VW126, the prototype, seen in flight. *A V Roe*

and all Lincolns. Power was offered by four of the new Rolls-Royce Griffon (Mk 57As inboard and Mk 57 outboard) engines with six-blade contra-rotating Rotol propellers that had initially been tested on Lancastrians VM704 and VM728. There were a number of problems relating to the rather complex mechanical nature of contra-rotating propellers and those on the 696 were developed by de Havilland at Lostock. It would not have succeeded at all if had it not been for the input of Rolls-Royce and most certainly that of 'Jimmy' Martin of Martin-Baker who solved all the problems related to lubrication of the hub end, which saw the most friction and heat, with his wrap-around translation unit. When the aircraft is in flight it is necessary every few hours to throttle back each engine and in so doing this allows oil to be sucked into the translation unit thus reducing the friction and heat produced at the hub end. This aircraft, known to Avro as the Type 696, became the Shackleton GR Mk I when it was adopted for service. The name Shackleton was actually chosen by Roy Chadwick, Avro's Chief Designer, in honour of famed voyager and explorer Sir Ernest Shackleton. Mary, Chadwick's wife, was a direct descendant of Ernest Shackleton, which explains why it is doubly apt, but you also have to remember that Roy Chadwick designed the Avro Antarctic seaplane for Sir Ernest Shackelton's last expedition in 1921, so he obviously had a great affinity with the man.

Initially all the performance data given by Avro in relation to the Type 696 was based on the use of the Rolls-Royce Merlin 85, and the Griffon was only considered as an alternative. The initial performance data projected a top speed of 300mph at 18,300ft, an absolute ceiling of 27,000ft and a service ceiling of 25,700ft, a maximum range of 3,800 miles at an average speed of 200mph and a gross weight of 82,000lb. It can been seen in comparison with the data of the later productions marks, that although pretty impressive for 1946, these figures did not tell the true story of the designs potential. Initial tests with the Griffon engine were conducted at the Rolls-Royce facility at Hucknall, Derbyshire. Here two modified Lancastrian II airliners were fitted with the Griffons inboard while retaining their original Merlins outboard.

The Prototype

Avro received the first contract, No.6/ACFT/ 10787/ CB6(a) dated the 28th May 1947, and it called for three prototype aircraft to be built. These airframes were allocated serial numbers VW126, VW131 and VW135 on the 17th July. The first of these prototypes, VW126, was rolled-out at Avro's Woodford facility and its structure and equipment belied its bomber ancestry. The aircraft featured Boulton Paul Type L barbettes on either side of the nose and a dorsal Bristol Type B.17 power-operated turret, both of which housed Hispano 20mm cannon. The 'cheek' barbettes offered 45° of vertical transverse, but no form of lateral movement. The type also had a Boulton Paul Type D turret in the tail housing two 0.50in Browning machine guns. There was a transparency under the nose to house a radar, a Flight Refuelling Ltd in-flight refuelling pick-up under the tail and a capacious bomb bay that was designed to hold up to 20,000lb of equipment and ordnance, including anti-submarine bombs, depth charges, sonobuoys, mines and marine markers. The radar system installed was the ASV Mk 13, which had an effective range, in good weather conditions, of 40 miles. This was greatly reduced if the weather was poor however. The 3ft Type 85 scanner was housed under the nose, although this was not the ideal location as it did not allow the full 360° of rotation. The crew consisted of ten men comprising two pilots, two navigators, a flight engineer and five other crew members to undertake the varied tasks dictated by the types role (these five crew members were gunners and signallers who also covered the roles of observation, and bomb-aiming. The design drew on experience with previous maritime designs and included such things as a galley and rest bunks for the crew. Johnny Baker from Avro had undertaken a fact-finding tour of Coastal Command units and having seen the Lancaster GR Mk 3s of Nos. 37 & 38 Squadron on Malta he had realised the need for proper crew facilities on a type that would be operating for very long periods. The navigators and electronics-operator were located side-by-side parallel with the fuselage side and the radio operator was positioned facing forward just aft of the electronics operators station. The large diameter fuselage was well sound-proofed, a very necessary inclusion when you consider the length of time that the type would be on one mission, and its size allows crew members to walk upright for its entire length down one side behind the seats of the other crew members. Crew access was via a door on the starboard side of the rear fuselage and a camera was located under the tail for both reconnaissance and to allow assessment of weapon delivery etc.

Even with the tragic death of Chief Designer Roy Chadwick in the crash of the Tudor Mk II prototype on the 23rd August 1947, the first prototype Shackleton was undergoing taxi trials at Woodford by mid-February 1949. The first flight of the Shackleton prototype VW126 took place on the 9th March 1949, with J.H. 'Jimmy' Orrel, Avro's Chief Test Pilot at the controls. He was assisted in this flight by Mr A. Blake as flight engineer and Mr S.E. Esler as co-pilot; 'Red' Esler was later killed in the crash of the Avro 707 prototype VX790. Initially Jimmy had not been impressed with the amount of movement he needed to make the rudders effective on the ground while taxying, so he had delayed the take-off and gone back to the hangar where some 'tape and glue' trimmers were added to the rudders. Satisfied with revised operation of the rudders he took off after a 14 second run and flew round Woodford for 33 minutes. On landing some more 'modifications' were made to the rudders and he then made another flight of 45 minutes. His assessment of the type was summed up when after the second flight he commented that 'we all knew that it was going to be a good aircraft from the start, it had the 'Chadwick Stamp' all over it'.

The aircraft moved to A&AEE Boscombe Down on the 21st April to start its manufacturer's acceptance trials and by using the staff of B Squadron at Boscombe Down the Avro crew was able to reduce the time required to complete the Ministry of Supply trials that would have followed, as many of them would have only mirrored the work undertaken by the Avro team anyway. The aircraft then moved to de Havilland at Hatfield in June to undertake strain-gauge tests on the contra-rotating propeller system, returning to Woodford in July. The strain-gauge test had not been successful initially, so VW126 returned to Hatfield around Christmas 1949 after modification.

The second prototype, VW131, made its maiden flight on the 2nd September 1949, with Johnny Baker at the controls. It was presented to the public for the first time at the SBAC show at Farnborough on the 6th September and was once again flown by Johnny Baker. For the SBAC show this aircraft did have the 20mm cannon installed in the nose and tail, but this was just for show, as it had already been decided that the 'cheeks' barbettes and tail guns were to be dispensed with on production aircraft. This machine was also the first to have guns fitted in the dorsal turret and it joined the first prototype at Boscombe Down on the 12th February 1950 where the main type trials took place throughout July. Production trials with the

20mm Hispano cannon in the dorsal turret were undertaken by VP263 at A&AEE Boscombe Down from the 20th July 1951. During the Autumn of 1950 VW131 went to Khartoum to undertake tropical trials, although it suffered a bird strike to the nose-mounted radome on its flight home. Neither VM131 nor VM135 had the in-flight refuelling receptical fitted to the first prototype. This system had seen the pick-up located in a fairing on the port lower edge of the rear fuselage on VM126, although no flight testing was ever undertaken with this system and it was soon dispensed with by Avro.

The third prototype VW135 was modified before completion to bring it up to production standards and as a result did not fly for the first time until the 29th March 1950. This machine joined the other two at Boscombe Down in July 1950. It was also tested by RAE Farnborough and was used to give type clearance for such systems as sonobuoy and flame-float launching, as well as service acceptance of the camera installation. This aircraft remained in use and undertook various roles until March 1954.

All three prototypes remained in use for various trials and development work. VM126 was used in February 1950 for cabin-heating and noise-suppression trials, with a Coastal Command crew seconded to Woodford for this purpose. Later it went into the shop at Woodford to become the aerodynamic prototype for the Mk 2, while VM131 and VM135 continued development trials for the Mk 1.

Because the type had initially been ordered off the drawing board, it was not long before Avro were actively looking at a revised and improved version to replace

Nice overall view of the Shackletons during the Royal Review in 1953. Either side of the Shackletons are Handley Page Hastings and Boeing Washingtons.

it even before it had reached squadron service. Production machines were needed immediately and the first one, VP254, flew for the first time on the 28th March 1950, the day before the third prototype. Avro had actually received contract No.6/ACFT/ 6062/ CB6(a) in March 1946, and it called for the production of thirty Shackleton Mk 1s before the contract for the three prototypes had even been written! The second production machine, VP255, flew for the first time on the 30th May and was present in the static park at the RAF Display at Farnborough on the 7th and 8th July. The aircraft then undertook a series of demonstration flights at the various Coastal Command bases identified for its potential use; these included St. Eval, Kinloss, Leuchars and Ballykelly during July and August. This machine moved to undertake Bristol Type B.17 dorsal turret trials at Boscombe Down in November and in August 1952 it was modified by a party from Avro at No.38 MU Llandow, South Glamorgan to prepare it for service use by Coastal Command, its use as a trials aircraft being at an end. The third production version to actually fly was VP257, which took to the air on the 28th August and this is the machine that Johnny Baker flew during the SBAC display at Farnborough in 1950. This machine then moved to the Central Service Department Establishment at RAF Wittering on the 13th December and remained there for four months. It left Wittering on the 13th April 1951 and moved to No.38 MU. VP256, which flew for the first time on the 18th September, undertook Pilot's Notes preparations at RAF Manby from the 28th September to the 5th November 1951. This aircraft then went to be updated at No.38 MU and was finally allocated to No.224 Squadron at Aldergrove on the 30th August 1952. Next to fly were VP258 on the 13th October 1951 and VP259 on the 24th October 1951. All of these machines were designated MR Mk 1, as their role had now been revised to Maritime Reconnaissance, instead of the General Reconnaissance (GR) designation initially used. In all twenty-nine MR Mk 1s were built and were allocated serials in the VP254 to 268 and VP281 to 294 range. It should be noted that the first production machine was actually provisionally allocated serial number VP253, but it was cancelled before construction commenced because the third prototype was so close to completion and this aircraft was brought up to production Mk 1 specification instead. The first of this second batch, VP281, flew for the first time on the 24th April 1951. When the final aircraft from this batch, VP294, flew on the 18th July 1951 it marked the end of Mk 1 production.

Into Squadron Service

With the formation of NATO in 1949 the initial plans to use the Shackleton as the sole land-based aircraft of Coastal Command in co-operation with the Sunderland flying boats had to be revised. The UKs commitment to NATO gave it a huge sea area to cover,

Neptunes of No.236 OCU and their crew line up not long after the type arrived.

P2V-5 Neptunes are collected at the Lockheed plant at Burbank by RAF crews.

The Commanding Officer of No.236 OCU when the Neptunes arrived was Sqn Ldr V R Cooney.
RAF

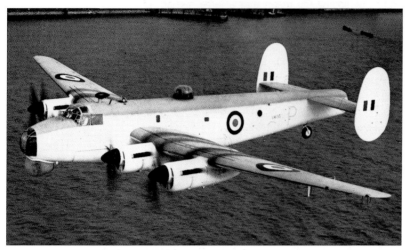

Starting its service with the US Navy as P2V-5 Neptune 51-15938, WX516 was redesignated as a Neptune MR Mk 1 WX516 and served with 210 Squadron at Topcliffe.

VW135 was the third prototype Shackleton built. After trials with the A&AEE and RAE was broken up for spares in April 1954.
Both via Martyn Chorlton

and this was way beyond the operational capacity of the aircraft then in service with Coastal Command. As a result the need to acquire a suitable type in quantity quickly was paramount. The British Government had discussions with the American Government and as a result fifty-two Lockheed P2V-5 Neptunes were loaned under the Mutual Defence Aid Pact. The type was to be a stop-gap measure until sufficient Shackletons could be supplied and they were to be used to patrol the North Sea and coastal waters. The main Neptune base was Topcliffe with crew conversion training taking place at Kinloss. A Neptune squadron would also be based at Kinloss for northern waters patrols and in all Nos.36, 203, 210 and 217 operated the Neptune from January 1952 until 1957 when all the remaining airframes were returned to the USA.

Initial Shackleton crew conversion was undertaken at Kinloss, and it was intended that squadrons would be based along the western seaboard at St. Eval in Cornwall, Aldergrove and Ballykelly in Northern Ireland and on Gibraltar. No.120 Squadron was to be the first to equip with the Shackleton, and they started to receive airframes from the initial production batch in April 1951. The first was VP258, which arrived on the 3rd April 1951 and it was quickly followed by four more (VP259 to VP262) from the initial production batch. The squadron was temporarily based at Kinloss, which was an ex-RNAS base that had been modernised by the Air Ministry 'Works and Bricks' Department prior to their arrival. Here it carried out service trials in liaison with No.236 Operational Conversion Unit which was also forming on the base. No.236 OCU was formed at

Kinloss on the 31st July 1947 by the redesignation of No.6 OTU which had acted as a twin-engine conversion unit using the Mosquito and Beaufighter during WWII. The OCU got its first Shackleton (VP264) on the 31st May 1951, where it was coded C•Z and by the end of the year it had a compliment of twelve aircraft for service training. The unit was redesignated the Maritime Operational Training Unit (MOTU) when it combined with No.1 MRS (Maritime Reconnaissance School) on the 1st October 1956, by which time it had used no fewer than twenty-two different aircraft. The Air-Sea Warfare Development Unit (ASWDU) was initially based at Ballykelly, having moved up from Thorney Island on the 27th March 1947. This unit received its first MR Mk 1 (VP261) on the 27th April 1951, although this aircraft only stayed with the unit for two weeks. The ASWDU moved to St. Mawgan on the 10th May 1951 and received MR Mk 1 VP282 on the 14th June 1951. Many sources state that the ASWDU received the next three production airframes after the initial batch of five had gone to No.120 Squadron for weapon development and installation trials, however it would seem that this is an error. No.224 Squadron on Gibraltar started to receive the type in July, with VP283 and VP287 being the first. The Shackleton replaced the ageing Handley Page Halifax Met Mk 6 that the squadron had previously operated and it had completely converted by October that year. No.220 reformed at Kinloss on the 24th September 1951 and received VP294 as its first Shackleton that same day. In November the squadron moved to St. Eval, which had been inaugurated in the autumn. No.269 Squadron reformed with the type at St. Eval in January 1952 and on the 14th March moved to Ballykelly in Northern Ireland. The Joint Anti-Submarine School situated at Ballykelly received its first Shackleton, MR Mk 1A WB849, on the 18th March 1952. This aircraft was followed by WB850 on the 26th March and WB851 on the 31st March.

The first squadron to use the type, No.120, completed service trials and moved to its permanent base at Aldergrove in Northern Ireland in April 1952. This was followed by the reforming of No.240 Squadron from a nucleus of No.120 Squadron crews at Aldergrove on the 1st May 1951 and the unit took on charge VP255 the same day, before moving to Ballykelly on the 6th June. No.42 Squadron was reactivated at St. Eval on the 28th June 1952. There first aircraft, MR Mk 1A WG509, had actually arrived there in March before the unit was officially re-formed. Finally No.206 Squadron was formed from a nucleus of No.224 Squadron crews at St. Eval on the 27th September 1952, receiving MR Mk 1A WB833 on the 11th October. By this stage there were seven Shackleton squadrons and these were supplemented by the Neptune squadrons that were also starting to become operational at this time. One of the last units to operate the MR Mk 1A was No.205, who received WB854 on the 14th July 1958.

The Mk 1A

By late 1951 Shackletons from the next (WB-series) production batch underwent a conversion which allowed the inner and outer engines to be inter-changeable. The outer nacelles were thus widened to accept the Rolls-Royce Griffon 57A engines. Fifty-eight more airframes were ordered in two batches to this revised specification and in this form they were designated the MR Mk 1A. All existing MR Mk 1s were later updated to Mk 1A standard during the 1955-6 period.

The first MR Mk 1A, WB818, made its maiden flight on the 1st August 1951. It then went to No.38 MU and then on to St. Mawgan before being issued in January 1952 to No.269 Squadron on Gibraltar. WB819, the second MR Mk 1A, first flew on the 2nd August 1951, but managed to beat WB818 to Gibraltar by arriving there on the 18th October 1951. By this stage it should be noted that the ample sound-proofing seen in the prototypes was not present in the production machines, the noise levels inside the MR Mk 1 and 1As were much higher as a result. Although the crews enjoyed the extra space inside the new large diameter fuselage, the higher levels of noise due to the lack of sound-proofing and the fact that the fuselage interior remained painted in matt black meant that it was a very depressing and tiring environment in which to work.

Final Service

1958 saw the final phasing out of the Sunderland Mk 5, when No.205 Squadron, previously at Seletar, Singapore, moved to Changi and started to re-equip with the Shackleton Mk 1A during May of that year. They completed this conversion in May, with their last Sunderland operational flight taking place on the 15th May that year and this marked the final standardisa-tion with the Shackleton in the long-range maritime reconnaissance role. The process of reorganisation and re-equipping did mean that a large number of units were disbanded and in an attempt to keep some famous squadrons alive a lot of renumbering took place. No. 202 became No.201 (10th October 1959), No.240 became No.203 (1st November 1959) and No.269 became No.210 (1st December 1959). At the time of this renumbering No.269 Squadron also re-equipped with the Mk 2, leaving No. 205 operating the Mk 1A, although they were joined by No.204 Squadron during the period May 1958 to May 1959 when they also operated the Mk 1A. All remaining Mk 1A airframes were either scheduled for conversion to T Mk 4, or they were disposed of either as ground instructional airframes, for fire training or sold as scrap from the MUs at Aldergrove or Shawbury. No.205 Squadron did not finally re-equip with the Mk 2 until February 1962 and it was only then that the last Mk 1A was finally withdrawn from service.

MR Mk 1, WB822 only ever saw service with No.236 OCU/MOTU before being converted to T Mk 4 standard.
via Martyn Chorlton

Right and centre right: Shackelton prototype VW131 seen on Malta during tropical trials. *Both Malta Aviation Museum Foundation*

MR Mk 1, WB848 seen while serving with No.236 OCU. It only ever served with this unit, in both No.236 OCU and MOTU guise, before going to No.240 Squadron and eventually being scrapped in October 1963.

Lovely official shot of an unidentified MR Mk 1 of No.269 Squadron.

Close-up of the nose of the prototype VW126. It had the 'cheek' barbettes, which were not adopted for production aircraft.

This manufacturer's image shows the in-flight refuelling pick-up on the port lower edge of the rear fuselage on VW126. This was only ever installed on this machine, as the next two prototypes and all production aircraft deleted this installation.
Both A V Roe

Third production MR Mk 1 VP256.

Centre left: MR Mk 1 VP257 of No.220 Squadron over London during the Coronation flypast on the 5th June 1952. *Christopher Ewing*

Centre right: The wrecked Shackleton is annotated 'Gibraltar Flypast error', and may relate to a day in the 1950s when a misunderstanding (!) left an aircraft on a low level flypast with too many engines feathered to remain airborne. However, the props don't look to be feathered, perhaps they were unfeathered too late, or the accident may be the result of the undercarriage problems suffered by early Shacks? A mystery after all these years. *Laurie Bruce*

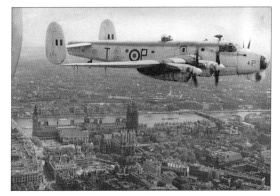

MR Mk 1As of No.269 Squadron at Ballykelly in March 1958.

MR Mk 1s and an early production MR Mk 2, at No.236 OCU, Kinloss. *Both Barry Jones Collection*

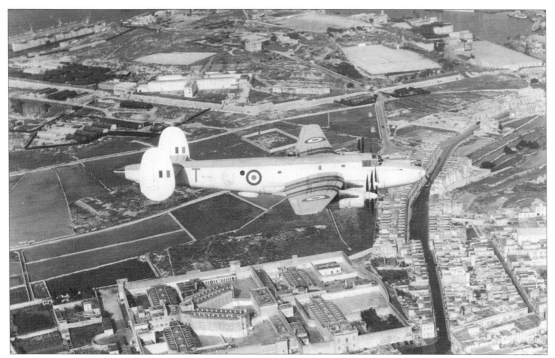

MR Mk 2, WL794 operated by No.38 Squadron off Malta and is seen here flying over the Grand Harbour area in 1954. It was lost in a crash off Gozo on the 12th February 1954.
R J Caruana

MR Mk 1, VP287 whilst in service with No.269 Squadron during 1957. This machine later served with No.240 Squadron before going to 23 MU at Aldergrove. Finally being scrapped in October 1963.

MR Mk 1, WB822 of No.236 OCU. The mid-upper Bristol Type B.17 turret is prominent in this shot. *via Martyn Chorlton*

Chapter 3: **IMPROVING THE TYPE**

The Mk 2

This evolution of the basic Mk I design was given the go-ahead with the release of Issue 2 of Specification R.5/46 on the 3rd July 1950. To meet this requirement, VW126 was rebuilt at Woodford during the winter of 1950-1 to incorporate all the aerodynamic elements of this new version. The nose was lengthened forward of the windscreen line and had a mock-up gun turret installed with a look-out position atop it. The gun position contained two 20mm Hispano cannon, remotely controlled from the upper look-out position. To understand why this was installed you should appreciate that the Shackleton was being considered to undertake the role of colonial policing and as a result, although a turret had little use in anti-submarine work (deck guns on submarines having disappeared from post-war designs), it would be useful in the policing role. The rear fuselage was lengthened and tapered to a cone. The original single tailwheel unit was replaced with a retractable twin wheel unit and toe brakes and lockable rudders were also installed for better ground handling as the Mk 1s had suffered considerably when taxying in a crosswind. The radar was relocated ventrally, aft of the wing, in a two-stage 'dustbin' although this was just a dummy unit at this stage. The bird strike suffered on the second prototype Mk 1 (VW131) when it returned from tropical trials had highlighted that a chin-mount-

ed radar was highly susceptible to damage. In this new position at last the full 360° rotation of the scanner could be achieved and thus the radar was far more effective in a tactical role. In this form VW126 flew for the first time on the 19th July 1951 and it moved on to A&AEE Boscombe Down for trials on the 25th July.

The first true MR Mk 2 prototype (VW126 was just an 'aerodynamic' prototype, to illustrate the 'look' of the new design) was WB833, which was a production MR Mk 1A that was taken off the production line and rebuilt to fully meet the Mk 2 specification. The mock-up nose turret seen in VW126 was replaced in WB833 with a Boulton Paul Type N unit fitted with two 20mm Hispano cannon and remotely controlled from the upper look-out station. Under this turret a prone bomb-aimer's station was also added and it featured an optically flat clear viewing panel. Another prone look-out station was also incorporated in the tail cone, which was fitted with a transparent cover. The radome, being under the rear fuselage, was a semi-retractable one to allow clearance with the ground and was fully operational in this airframe. In its Mk 2 form WB833 flew for the first time on the 17th June 1952. On the 23rd July it moved to Boscombe Down where it undertook engineering, performance, navigational and radio trials for the next four months. The aircraft moved to RAE Farnborough on the 2nd December for radio compass trials and after completing this it moved back to

MR Mk 2, WL794 was only ever operated by No.38 Squadron off Malta and crashed into the sea off Gozo on the 12th February 1954. It is seen here in happier times flying over Grand Harbour in early 1954. *R J Caruana*

Top: Shackleton VW126 is seen here as the aerodynamic prototype for the MR Mk 2 series.

Above: Shackleton MR Mk 2 WL796 was used for trials with the Airborne Lifeboat Mk 3. It is seen here at Farnborough with it installed, although it was never actually dropped from the Shackleton or adopted for use with the type by the RAF. *Both A V Roe*

Woodford on the 29th January 1953 and started manufacturers performance trials.

These trials were a success, and thus the last ten airframes off the Mk 1A production batch (WG530-533 and WG553-558) were all completed to Mk 2 standard. The first three production Mk 2s were completed in September 1952 and were then loaned to the Ministry of Aviation (which the Ministry of Supply had become) for trials, with WG530 going to A&AEE Boscombe Down on 25th September. WG533 became the first Mk 2 to be delivered to the RAF when it went to RAF Manby mainly for the compilation of the Pilot's Notes for the type and, at a later stage, tropical trials while WG532 ended up at the ASWDU for trials with underwing rockets in January 1953. These proved unsuccessful however and rockets were never adopted for use with the type.

Seventy MR Mk 2s were built, covered in three contracts, and carried serials WB833, WG530 to 533 (originally ordered as MR Mk 1As), WG553 to 558 (originally ordered as MR Mk 1As), WL737 to 759 (from the first real MR Mk 2 contract), WL785 to 801 (also from the first real MR Mk 2 contract) and WR951 to 969.

Into Service

The first batch of MR Mk 2s (WG530 to 533 and WG553 to 558) were not allocated to squadrons by the time that the first MR Mk 2 from the second production batch (WL737) made its first flight on the 17th November 1952. Most of these initial machines were used for service trials associated with the adoption of the type for service life, although eventually most were allocated to operational squadrons. The exception to these were WG557 and 558, which both went straight to No.260 Squadron after having been accepted for service at No.23 MU.

The first squadron to adopt the MR Mk 2 was No.206 Squadron based at St. Eval, when they received WL737 on the 24th March 1953. This was followed by No.240 at Ballykelly who received WL738 and WL739 on the 31st March. Later St. Eval based No.46 Squadron also received the type and it also equipped No.269 on Gibraltar and No.120 at Aldergrove. No.42 Squadron sent three Mk 1s and three Mk 2s on a tour of Ceylon and South Africa on the 13th April 1953 and in July their remaining machines were involved in the Royal Review at RAF Odiham.

The final two squadrons to equip with the Mk 2 were No 204 and 228. The former reformed at Ballykelly on 1st January 1954, while No.228 reformed at St. Mawgan during June. Problems with operating the Mk 1 and 1A alongside the newer Mk 2 soon became apparent and so the decision was taken to standardise each squadron to just one type. Although the three marks may sound similar, they were all very different machines and as a result the amount of spares and equipment needed to support all of them led to ratio-

nalisation. As a result in the Summer of 1954 Nos.120, 204, 206, 220, 240 and 269 all relinquished their Mk 2s, while Nos. 42, 210 and 224 Squadrons totally equipped with the Mk 2. Once all this was sorted there were eleven Shackleton squadrons supplemented by four Neptune and three Sunderland squadrons covering the Atlantic from the Arctic to the Equator as well as the whole of the Mediterranean.

The second batch from the second order commenced with WL785, which made its first flight on the 10th May 1953. This was followed by WL786 on the 12th May, WL787 on the 18th and WL788 on the 4th June. With more airframes being available re-equipping of other units could now take place and in August and September 1953 Nos 37 and 38 Squadrons on Malta received the Mk 2 to replace their Lancaster GR Mk 3s. Their last machine, RF273 holds the accolade of being the last operational Lancaster of Coastal Command when it left the unit on the 3rd February 1954. No.37 Squadron at Luqa received WL785, 786 and 788, while No.38 Squadron, also at Luqa, got WL787.

The final batch of Mk 2s to be built was to contract 6/ACFT/6408/CB(6a) dated 8th February 1951. Initially this order was for forty aircraft, but in the end just nineteen were made as complaints were still common from service crews about the poor crew environment and noise and vibration levels. The remaining twenty-one aircraft in this order were therefore eventually built as MR Mk 3s. This final batch of Mk 2s was allocated serials in the WR951 to WR969 range and WR951 had its first flight on the 20th October 1953. This machine also became the first Mk 2 from this final batch to go to an operational unit when it was accepted by No.204 Squadron on the 4th January 1954. This was followed on the 10th February 1954 with the issue of WR952 to No.206 Squadron at St. Eval. WR953 went to No.228 Squadron from No.38 MU, WR954 joined No.37 Squadron at Luqa and WR955 was issued to No.120 Squadron on the 8th April 1954. The next seven aircraft, WR956 to 962, had gone to the reformed No.228 Squadron at St. Eval from the 1st January 1954. The last aircraft in this final Mk 2 production batch, WR969 also held the record for operating with the most number of units. It served with Nos.38, 204, 205, 210, 220, 224 and 228 Squadrons, as well as the MOTU and JASS during its service career from December 1954 through to September 1971 (it had been converted to a T Mk 2 from 1967 to 1970 and then converted back again).

Trials Work

Trials with new equipment continued in the Mk 2, with WL789 being fitted with a MAD (Magnetic Anomaly Detector) boom in an extended tail. This work was carried out by a work party from Avro at No.38 MU Llandow from the 15th June 1953. Once converted the aircraft moved to St. Mawgan on the 10th

September and was taken on the strength of the ASWDU there on the 30th. Extensive trials were undertaken with the MAD boom over the next four and a half years, but the Shackleton proved unsuitable for this equipment due to the high resonance of the engines and contra-rotating propeller. The heavy nature of the types controls made it impossible for this type of detector to be used in an efficient tactical manner either. It has also been said by many that the reason that MAD did not work in a Shackleton was simply because the Shackleton was a magnetic anomaly all on its own! WL789 was converted back to standard configuration by the ASWDU and returned to No.49 MU at Colerne where it was updated to Phase 1 standards before going to No.224 Squadron on Gibraltar.

Trials using a Neptune were also undertaken in relation to rocket-fired 'Glow Worm' illumination flares ini-

Two MR Mk 2s from No.38 Squadron in flight. Those involved (probably) include WL759 'V' (foreground) and WL755 'U' (background).
Christopher Ewing

tially intended for the Shackleton. This type of unit was found to be very hazardous to use and although the Mk 3 was built with the wiring in the wings for carrying such rockets, they were never adopted by Coastal Command. The South African Air Force (SAAF) did adopt the system and all their MR Mk 3s were fitted out to carry the rockets.

WL796 was used for trials with a Mk 3 Airborne Lifeboat and although the aircraft appeared with this fitted at the SBAC display at Farnborough in 1953, it never actually dropped it and the lifeboat was never taken into service use with the Shackleton as Lindholme Gear was far more effective in an SAR role.

Goodwill Ambassador

The Shackleton was a very rugged aircraft, that could stand the rigours of operating at sea for long periods and it had the added benefit of being able to support itself while away from base. The large fuselage made it easy to carry ground crew and spares, while tools and equipment could be loaded into panniers and carried in the large bomb bay. As a result of this the type was

to undertake many detachments to NATO member countries for exercise and goodwill tours.

The first such tour was undertaken by MR Mk 1As of No.220 Squadron, who visited Ceylon in February 1952. No.224 also visited the island in August of the same year with their MR Mk 1As. The Mk 2s of No.42 Squadron also visited Ceylon in 1953 and sent a detachment to South Africa; the result of this being the only foreign export of the type, when the SAAF ordered the type a number of years later. Four Mk 1 & 1A aircraft of No.206 Squadron did a seven week Commonwealth Goodwill Tour from 12th August 1954 with exercises in Ceylon and visits to New Zealand and Fiji. No.204 Squadron sent four of the Mk 2s to Durban, South Africa in June 1955 for exercises (Exercise Durbex II) with the SAAF and Army, flying via the Gold Coast for a goodwill visit on the way back. In July 1960 four MR Mk 2s of No.204 Squadron flew to Singapore via Kindley Field (Bermuda), Palisadoes (Jamaica), Trinidad and Stanley Field (British Honduras). While at Trinidad the squadron proved the versatility of the type be taking troops from the local regiments (29 per aircraft) to Stanley Field for troop rotation.

A Time for Change

Although the Mk 1, 1A and 2 had all proved to be excellent maritime reconnaissance aircraft the conditions for the crews were not good, and as a result of complaints fatigue trials were undertaken by the RAE at Farnborough throughout 1953 and 1954. These trials were undertaken by the Institute of Aviation Medicine (IAM) and utilised crews from No.240 Squadron at Ballykelly. This extensive trial required the crews to undertake sixty hours of flying at night over a seven day period. Taking off between 6 and 6.30pm each day, the crews would continually fly through the night and land again around 7am the next morning. Each crew member was monitored throughout the flights to determine the effects on their abilities. It was discovered that each and every one of them were effected in one way or another. Most noticeable was the continual buzzing in their ears from the propellers and the physical loss of several pounds in weight after each sortie. All of these factors highlighted the need for revisions to the design and as a result, although more Shackletons were required to replace the ageing Sunderlands, production of the Mk 2 stopped with WR969 in September 1954. A complete redesign was called for, with a list of improvements in many areas plus full modernisation and this resulted in the Avro Type 716 which was to become the Shackleton MR Mk 3.

Above: WB833, the prototype MR Mk 2, drawn off the MR Mk 1 production line and rebuilt to MR Mk 2 standard, first flew on 17th June 1952. *Barry Jones Collection*

Right: Val DeMain's No.236 OCU crew at Kinloss on the 14th February 1952. From right to left they are Fred Coy, ? Evans, 'Chalky' White, ? Kavagnah, Val DeMain, 'Jock' Brown, 'Titch' Butler, Ken Cresswell, Bill Meichan and Steve Vassey. *Steve Vassey*

MR Mk 2, WL742, B-Z first flew on 23rd December 1952 and was sold for scrap in June 1968. *Barry Jones Collection*

WG531, the second production MR Mk 2, appeared at the 1952 SBAC Display at Farnborough. *Barry Jones Collection*

Image from No.205 Squadrons 50th Anniversary Parade at RAF Changi, Singapore in December 1966.
Laurie Bruce

MR Mk 2, WR957, 'U' of No.228 Squadron, approaching a rocky UK coastline.

Below: WL742, 'Z' was the sixth production MR Mk 2, seen here while serving with No.206 Squadron in January 1954.
Both Barry Jones Collection

Below: This photograph shows an RCAF Argus, an ASWDU Shackleton and a USN P3 and it was taken at NAS Key West, Florida in 1966, when the ASWDU attended the first annual conference on anti-submarine warfare development. The Argus and P3 are from the RCAF and USN equivalent units.

Top left: Forty-eight propeller blades thrash the air over a vast expanse of ocean.
Barry Jones Collection

Far left upper: No.220 Sqn aircraft 'N' at low level over a submarine (probably not British) 22nd March 1953. Photograph was taken from the co-pilot's seat using a K20 camera, which was the large, rather clumsy but efficient hand-held camera in use on Shackletons. The aircraft was being flown by Sqn Ldr Laband.
Laurie Bruce

Left: This nine aircraft formation shows the Maritime Operational Training Unit (MOTU) leaving RAF Kinloss for RAF St. Mawgan on 2nd July1965. *Laurie Bruce*

Far left: This photograph,taken on 1st June 1969, at 77North shows the British North Greenland Expedition Camp. It was taken on a 17 hour 25 minute sortie flown to test the effectiveness of newly produced radio propagation tables when flying at low-level at high latitudes. On that day the crew took off at 0815 from RAF Kinloss for RAF Lossiemouth, where they were to refuel and use the long runway, but had to return after losing an exhaust stub on take-off. The stub was replaced and they took off again at 1000 for a 15 minute flight to Lossiemouth. After fuelling up we took off from Lossiemouth at 1155, then landed back at Kinloss at 0520 the following morning! With briefing and debriefing it was a 24 hour day. On 15th June we flew a 17 hour 10 minute flight to Bear Island and return on a similar mission, again after a short hop to Lossiemouth to fuel up.
Laurie Bruce

Chapter 4: **THIRD TIME LUCKY!**

The Mk 3

The Mk 2 did not really address any of the shortcomings of the Mk 1 or Mk 1A. It is true to say that some aspects were changed, like the location of the radar scanner, but basically the MR Mk 2 was just a 'makeover' of the first version and as such was little more than a continuation of the earlier version that just looked different!

The third revision (Issue 3) of Specification R.5/46 was issued on the 18th November 1953. Actual work on the new design did not start until early 1954 however. The revised aircraft featured tricycle undercarriage with twin nose and main wheels and a hydraulically operated Dunlop Maxaret braking system, all of which improved airfield performance, especially in a crosswind. The revised nose was more sloped and although the type retained the 20mm cannon, the sighting device was updated. The new nose undercarriage resulted in the need to shorten the bomb bay, as the nose wheels retracted backwards. There was also need to shorten the bomb bay due to

the repositioning of the crew access hatch, although the old crew access door in the starboard side of the rear fuselage was retained and now used as an emergency exit. The fuselage also had a retractable tail bumper fitted just aft of a new tactical camera in the lower rear fuselage. The wing changed with modified ailerons of greater chord and the addition of permanent wing tip tanks. The latter held 250 Imperial Gallons each thus increasing the fuel capacity to 4,716 Imperial Gallons and as a result a fuel jettison system was also installed. It had been hoped that this increased fuel capacity would increase the operational range of the Mk 3, but it increased the all-up weight so much that the Mk 3 actually had a slightly shorter range than the Mk 2. The dorsal turret was deleted and the resulting space was used to install full cooking facilities for the crew. This area was a risk, due to the cooking and associated sources of combustion, so it was sealed off fore and aft of the wing rear spar by bulkheads. The crew could pass through this area on the starboard side via doors in the bulkheads. A

A Charles E Brown image of prototype MR Mk 3, WR970 on an early flight. The lines on the upper wings denotes walkways for servicing crews and run on top of the main spars.

WR970, the prototype MR Mk 3, first flew on 2nd September 1955 and was used by the A&AEE for handling trials before it crashed in Derbyshire on 7th December 1956.

The 1955 SBAC Display showed WR970, the seventh production MR Mk 3, in the static park before it joined No.220 Squadron.

WR970 landing at Farnborough (SBAC) in 1955. All *Barry Jones Collection*

A very posed RAF official photograph of a groundcrewman 'cleaning' his aircraft!

Boscombe Down on the 7th September 1956. In the company of a team from Avro the aircraft was to undertake development trials, however it did not have an easy time during these trials as its stall characteristics were not acceptable and the aircraft was returned to Avro on the 28th November 1956 for an improved stall warning device to be fitted before trials continued. The aircraft returned to Boscombe Down on the 7th September to resume trials work. Unfortunately during a test flight from Boscombe Down on the 7th December the aircraft suffered a loss of control while the crew were investigating the stall characteristics and the aircraft became inverted in cloud. Little exact detail is known as flight recorders and other data systems did not exist at this time, but it is believed that although the crew managed to regain control the engines could not be restarted (the plugs having oiled up) and the windmilling effect of the contra-rotating propellers caused the aircraft to stall once again. This time the crew did not recover it and it hit the ground near Foolow Village in Derbyshire, with the loss of all four crew members (Sqn Ldr Jack Wales and three Avro technicians).

The second MR Mk 3, WR971, did not fly until 28th May 1956 and this machine was fitted with vee-shaped blocks on the leading edge of the inboard wing section to destroy the air flow on this inner section thus inducing a stall only at this point and thereby ensuring the entire wing only suffered the stall in a safe wings-level attitude.

Mk 3 Service Trials

With the loss of the prototype the service adoption of the type was put on hold. The second prototype, WR971, although initially held by the manufacturer, did start armament trials with the A&AEE on the 12th December. These were curtailed however and the aircraft went back to Woodford while the improved stall warning system was developed. Once the new system was perfected, the aircraft returned to A&AEE and underwent the rest of the armament trials and it was not until the 12th May 1958 that it returned to Woodford. The third production aircraft, WR972, flew for the first time on the 6th November 1956. It was retained by the Controller (Air) and used for various trials relating to photographic equipment and navigation radar. This aircraft later moved to the A&AEE and was eventually purchased by the Ministry of Aviation for use by the RAE at Farnborough. The fourth production aircraft, WR974, had its first flight on the 1st May 1957 and was used for tropical trials. The aircraft was prepared by Avro and delivered to A&AEE Boscombe Down on the 5th July. It was then flown out to Idris in Libya to undertake the trials. The aircraft returned to the UK in August 1957 and was flown into No.49 MU at RAF Colerne where a working party from Avro started to prepare it for winterisation trials. The aircraft was accepted once again by the A&AEE on the 20th September and was flown out to Canada for the start

revised perspex canopy was fitted that improved pilot visibility, achieved by dispensing with the framing seen in the previous examples. All electronic equipment was fitted to racks at roof level with drop-down doors and the whole interior was once again sound-proofed and lined with brown and cream Rexine artificial leather! Crew comfort on their long-haul missions was further improved with large comfortable chairs and the old exhaust stacks were replaced with sealed units that effectively and quietly deflected the exhaust gases under the wing. The design also took some account of potential future development in electronic equipment and so space for future development was built in from the start. From the outset the MR Mk 3 was designed to exclusively use the Griffon 58.

The prototype MR Mk 3, WR970, flew for the first time on the 2nd September 1955 and appeared at the SBAC event at Farnborough later in the month. The aircraft was actually a production aircraft from the main contract and as such was taken on the charge of the Controller (Air) while all service trials were conducted by the manufacturer. The aircraft completed trials with the manufacturer and moved to A&AEE

of the trials on the 8th October. The aircraft was in need of repairs prior to the trials commencing, so this was undertaken by The Fairey Aircraft Company of Canada, it then undertook the trials and returned to the UK on the 16th April 1958.

In Squadron Service

The first unit to re-equip with the MR Mk 3 was No.220 Squadron based at St. Mawgan when they got the sixth production aircraft, WR976, on the 30th August 1957. In December the unit was renumbered No.201 Squadron and they were joined by No.206 Squadron on the 14th January 1958, although this unit was only partially equipped with the Mk 3 due to slow delivery of the type from Avro. There were a few initial teething problems, mainly centred around engine fade caused by air entering the fuel system and hydraulic failures due to pipework of insufficient strength, but eventually the type was cleared for service use. No.120 at

Aldergrove and No.203 at Ballykelly re-equipped with the type in September and December 1958 respectively. The former squadron had been formed by renumbering No.240 Squadron and it received its first Mk 3, XF702 in November 1958. Initially the continuation order had been for fifty-two Mk 3s, but a defence review resulted in this being cut back to thirty-three with the second part of the order covering XF730 to XF734 being cancelled on the 14th February 1956. This cancellation did not effect XF730 though, as this was produced as a replacement to WR790 which has been lost in the fatal crash, and this replacement was delivered in June 1959. Thus the Mk 3 production run went from WR970 to 990 and WF700 to 711, with XF730 being a replacement for WR970.

XF700 was delivered to No.120 Squadron in September 1958, with XF701 arriving in October. XF702 went to No.203 Squadron as recounted earlier and this was joined by XF704 during the same month,

MR Mk 3 prototype WR970 is seen here at Woodford with the long tube installed in the nose to carry the yaw meter. The aircraft had just returned from the RAE and was sadly lost in a fatal crash three months after this photo was taken.

A nice image of MR Mk 3 prototype WR970 in flight. *Both A V Roe*

with XF703, XF705, XF706 and WR974 coming directly to the unit in December.

It was intended that another batch of MR Mk 3s would be produced and contract No. 6/ACFT/11106/CB6(a) dated July 1954 was issued to this effect. The contract called for thirteen new aircraft, allocated serials XG912 to XG924, but this order was cancelled on the 14th February 1956 before any were built.

Mk 3 in South Africa

The only Shackletons to be exported were those ordered by the SAAF. This country's interest had been kindled by the goodwill visit by No.42 Squadron in 1953 and further interest grew when No.204 Squadron came to the country for Exercise Durbex II in

1955. Having been made aware of the new MR Mk 3 and the intended Phase Programme of modifications, the South African Government felt that the type would be ideal to replace their ageing Sunderlands. In March 1954 Avro was issued with a contract for eight airframes from the initial batch of MR Mk 3s produced (construction numbers 1526 to 1533) and they were assigned SAAF serial numbers 1716 to 1723. All of these machines were built to Phase 1 standard.

In February 1957 forty personnel from No.35 Squadron, SAAF commanded by Commandant M.J. Uys, arrived at Woodford. Here they were familiarised with the aircraft. The first two Shackletons were officially handed over in a ceremony at Woodford on the 16th May 1957. These machines were temporarily

MR Mk 3 '1721' was ferried to South Africa in February 1958. One of eight aircraft ordered by the SAAF, '1721' retired in 1984 she is preserved at the SAAF Museum.
via Martyn Chorlton

Sunderland Mk III ML824 making its way up to Pembroke Dock on the 24th March 1961. Being an ex-No.201 Squadron aircraft, it was escorted in by WR975 and WR980 of that squadron. This image was taken from WR975 'P', as the aircraft in the background is WR980 'O'.
Author's collection

based at RAF St. Mawgan, arriving there on the 21st May, and then took part in a joint exercise with Coastal Command Shackletons over the North Sea. The first two aircraft (1716 and 1717) were involved in this work-up process for the next three months, then they were prepared for their flight to South Africa. The two aircraft started their five-day flight on the 13th August, arriving at Cape Town on the 18th. The next two aircraft, 1719 and 1720, did not make their way to South Africa until the 13th February 1958. The next two aircraft produced were 1721 and 1722 and they were joined at St. Mawgan by 1723 on the 13th February 1968. On the 14th February these three aircraft started their journey to South Africa, arriving at Ysterplaat AFB on the 26th February (1723 suffering a hydraulic failure en route and landing with no brakes, it ran off the end of the runway).

Initially it had been intended that the Shackleton would be based at Ysterplaat AFB, however this base did not have a long enough runway for the type to safely operate from. As a result all SAAF Shackletons were operated by No.35 Squadron from D.F.Malan airfield, Cape Town, while Ysterplaat AFB served as the maintenance and service base for the type.

The SAAF machines differed from their UK versions, although initially this only really consisted of the astrodome, which was placed well aft in comparison with Coastal Command machines. Three airframes were fitted to carry the SARO Mk 3 airborne lifeboat, but the SAAF, like the RAF, opted for Lindholme gear and the remaining airframes were never adapted to carry the lifeboat. The SAAF did opt for a fitment that the RAF had not gone ahead with though and this was the fitment of rocket rails under the outboard wing panels to carry Glow Worm illumination rockets. Much upgrading of the ECM fitment had also been carried out over the years but it is interesting to note however that the SAAF never opted to go for the Viper 203 jet engine installation seen in Mk 3s in the UK.

In service the Shackleton was used for a variety of roles. Initially they were used for 'border patrol' work over the 3,000 miles around the Kalahari Desert and Southern Rhodesia. These low-level flights upset the mass of big game in the area, and numerous complaints from the National Game Park Wardens lead to the flights being stopped.The type was also used to monitor Soviet shipping activities around the Cape of Good Hope, especially between the Indian Ocean and Atlantic and it also undertook fishery protection and SAR roles in the same areas. Although not a military role, the SAAF Shackletons also gave top cover for civil airliners of the South African Airways as they overflew the Indian Ocean on their way to Australia etc.

The SAAF suffered its first and only loss of a Shackleton on the 8th August 1963, when 1718 struck high ground before crashing into the Wemmershook mountain range near the town of Worcester. All of the crew were killed and the aircraft had amassed just 777 hours since being delivered to the SAAF six years previously. The seven remaining aircraft were updated to Phase 3 standard. This was undertaken by a working party from Hawker Siddeley, as Avro had, by this stage, been absorbed into this firm. Two aircraft were also re-sparred in their service life; 1716 and 1717. The former was done between March 1973 and April 1976, while the latter was done between September 1975 and October 1977. On the 22nd November 1977 1723 was grounded, having expended its fatigue life. Initially it was stored at Ysterplaat AFB, but it was eventually sold to Mr Vic De Villiers in a triple deal between him, the SA Airways Museum and the SAAF Museum. Mr Villiers mounted 1723 on the roof of his garage situated alongside the Johannesburg to Vereeniging highway and by 1994 this machine had been painted red and white overall and sported Coca Cola advertising! The next aircraft to be grounded was 1719 on the 24th April 1978 and it too was initially stored at Ysterplaat AFB before being moved to the Cape Town Waterfront complex in 1991 and finally being scrapped in the late 1990s. 1720 was next to reach the end of its fatigue life and it was grounded on the 10th March 1983 and this aircraft ended up outside the Warrant Officer's Club at Ysterplaat. 1717 was next to be grounded although this aircraft stayed airworthy until 1984 thanks to parts recovered from 1718 that had crashed in the Wemmershook Mountains back in 1963. This aircraft was also initially stored at Ysterplaat, before being dismantled and shipped by road and sea to its current resting place at the Natal Parks Board Museum.

The Shackleton was officially retired from SAAF service on the 23rd November 1984 and 1716, 1721 and 1722 made a flypast at D.F. Malan to mark the event. Two weeks later 1716 and 1721 were flown to the SAAF Museum complex at Swartkop, while 1722 was to remain in running order with No.35 Squadron. In 1991 this aircraft flew to Ysterplaat, which had by then become the second largest aircraft museum in South Africa. The aircraft remained in operational condition although as this book is being written we have learnt that it is the intention of the SAAF to finally ground this aircraft some time during 2005, thus marking the final chapter in the 48 years that the type has been in South Africa.

Of course many of you will know of the unsuccessful attempt to send an airworthy Shackleton to the UK in 1994 and we will cover this in more detail at the end of this book.

MR Mk 3, XF730, 'C', the last new-build Shackleton, belonged to No.206 Squadron when attending Wethersfield's Air Display on 28th May 1960.

No.38 Squadron Detachment to Ballykelly in May-June 1959.
Christopher Ewing

MR Mk 3, XF701 seen with No.206 Squadron in 1964 prior to being updated to Phase 3 standard as can be seen by the lack of Viper exhaust in the outer engine nacelle.
via Martyn Chorlton

Above: MR Mk 3, WR979 is pictured here after Phase 3 modification with the Kinloss Wing. It remained with this unit until put in store and then finally broken up at St Athan in 1971.

Left: MR Mk 3, WR976 seen in the colours of No.206 Squadron. This aircraft later went on to serve with the Kinloss Wing and No.201 Squadron and was sadly lost off Lands End in November 1967 *Both via Martyn Chorlton*

Left: Image from No.205 Squadrons 50th Anniversary Parade at RAF Changi, Singapore in December 1966. Parade state; Reviewing Officer Gp Capt GE Livock DFC AFC RAF (Rtd) (CO No.205 Squadron Jan 1929 – Feb 1931); Parade Commander Wg Cdr J.V.E.P. Carter; Adjutant Flt Lt L. Bruce [Self]; OC No.1 Flight, Sqn Ldr S.W.R.A. Key; OC No.2 Flight, Sqn Ldr G.J. Page; Standard Bearer Fg Off M.G. Christy; Standard Warrant Officer Master Signaller B. Collington; Escort to the Standard Sgt H.R. Parkinson & Sgt K.J. MacBrayne. *Laurie Bruce*

Chapter 5: **TRAINING, PEACE KEEPING AND ATOMIC TESTS**

The Shackleton Trainer

By 1956 the decision had been taken to standardise the Shackleton and plans were made to phase out the Lockheed Neptunes. The training system was also overhauled, as previously aircrew had spent two and a half months flying the Lancaster GR Mk 3 at RAF St. Mawgan before proceeding to No.236 OCU at Kinloss to be trained on their assigned aircraft. This was now all amalgamated into one large organisation entitled the Maritime Operational Training Unit (MOTU) at RAF Kinloss. The need for a trainer version of the Shackleton was now highlighted and as such a special training version of the Mk 1A was ordered as the T Mk 4. The prototype T Mk 4 was VP258 and although officially on the strength of No.120 Squadron it was actually in store. Therefore in May 1955 it was made ready and flown to Woodford for conversion. The conversion saw more radar stations being installed in the bunk area and this was made possible because the dorsal turrets on the Mk 1, 1A and 2 had all been systematically removed during the 1955/6 period. These machines also had dual-controls and were able to operate as both flight deck crew trainers as well as training the electronic equipment operators. The prototype was a Mk 1, so it had to have the modifications associated with the installation of the Griffon 58s made to the outboard engine nacelles as well. It went to A&AEE Boscombe Down for initial trials and then

moved back to Avro's Langar factory on the 3rd May 1957 to have ASV 21 installed, bringing it up to full T Mk 4 specification. VP258 was later updated to Phase 2 standard and after further acceptance trials with the A&AEE it was issued to the MOTU at Kinloss on the 12th January 1959 and remained there (with spells away at Bitteswell and No.49 MU) until it was sold to the Board of Trade and then moved to the Fire School at Stanstead on the 17th July 1968.

MR Mk 1s VP259 and VP293 were also converted to T Mk 4 standard, with the former converted at Woodford and the latter at Langar. VP259 was taken on by the MOTU on the 22nd July 1957, while VP259 was used for trials work for three years before going to the MOTU on the 1st March 1960. VP259 was unfortunately lost in a crash near Elgin on the 10th January 1958, but VP293 went on to be used for weapons and low-level TV trials by RAE Farnborough for many years until it was withdrawn in June 1975 and sold to the Strathallen Museum. Unfortunately it was ultimately scrapped, but the front fuselage was saved and is currently housed at Woodford on behalf of the Shackleton Association.

All of the remaining T Mk 4s were converted from MR Mk 1A airframes. The first was WB819, followed by WB820, WB822, WB826, WB831, WB832, WB837, WB844, WB845, WB847, WB849, WB858 and WG511. The last of these, WG511, left Langar as

Built as a Mk 1A, WB837 is seen here in T Mk 4 guise whilst with the MOTU. This is the only unit in which this aircraft served as a trainer and was eventually SOC as scrap in February 1969. *via Martyn Chorlton*

Mk 1A, WG525 seen here while with No.42 Squadron. It later went on to serve with Nos.220 and 205 Squadron before being damaged while on Gan. It was finally scrapped during August 1964. *via Martyn Chorlton*

a T Mk 4 in July 1957. All seventeen airframes converted to T Mk 4 configuration were operated by the MOTU at RAF Kinloss. The Lancaster GR Mk 3s were finally phased out there, with the last one (RF325) departing this base on the 15th October 1956. In July 1965 the MOTU moved south and was based at RAF St. Mawgan, where the remaining T Mk 4s were replaced in 1968 with the T Mk 2.

The T Mk 2 version was actually based on the Mk 2 Phase 3 aircraft and ten of them were converted at Langar in 1967. The first aircraft thus converted was WL739 a Phase 2 aircraft that was withdrawn from use by No.204 Squadron in September 1964. It flew to Langar where the 20mm cannon were removed, although most changes with the T Mk 2 were internal and not externally visible. The rest bunks and galley were removed and in this area a new master radar was installed, connected to a slave unit that was fitted in the place of the bunks. The aircraft was completed in this guise by the end of 1967 and was flown to the MOTU on the 3rd January 1968. Nine more airframes were also converted, starting with WL750 on the 14th December 1966 to WR967 on the 5th September 1966. This last aircraft did not remain in service long, as it was damaged at Kinloss on the 7th September 1972 and then converted to an AEW training aid, where it was allocated Instructional Airframe No.8398M and christened 'Dodo'! The airframes converted to T MK 2 were WG533, WG554, WG558, WL739, WL750, WL787, WR964, WR966, WR967 and WR969, although this does not identify the order in which the conversions took place. The last operational aircraft was WL787, which reached the end of its fatigue life on the 3rd January 1974 and was broken up by the fire fighting unit on site two months later.

Operations in Cyprus

The worsening situation with regard to Cyprus that existed between Greece and Turkey by the mid-1950s resulted in the Shackleton being called in to carry out anti-smuggling patrols to stop the illegal import of weapons to the island. The formation of the terrorist organisation EOKA, who were for union with Greece, lead to an escalation in the number of sabotage and bombing attacks. The inadequate number of British troops on the island was highlighted by the escalation. So in December 1955 the decision was taken to instigate the movement of British troops into the area under Exercise Encompass. From December 1955 through to the 24th January 1956 this exercise saw numerous Shackletons, from just about every unit that operated the type, being used as troop carriers

Each aircraft could carry thirty-three troops with all their equipment carried in panniers in the bomb bay.

Malta-based No.38 Squadron had been given the task of patrolling the entire coastline of Cyprus and it undertook its first sortie in this role on the 21st July 1955. They continued in this role for some four years, although in 1957 they were assisted for a while by No.42 Squadron. By 1959 the whole situation had eased in Cyprus and No.38 Squadron undertook their last sortie in the role on the 14th December 1959.

The Suez Crisis

The trials with the type as an emergency troop carrier proved very apt, because it was soon put to a full operation test as trouble flared once again in the Mediterranean. A British/Egyptian agreement had been signed in October 1954 and as a result of this British forces had been withdrawn from Egypt by June 1956. This was followed in February 1955 by the signing of the Baghdad Treaty between Iran and Turkey and in April by a similar agreement between Britain and Iran. By the time British troops were removed from Egypt the situation was such that keeping the previously signed agreements in force seemed unlikely. As a result the Baghdad Treaty was extended to include Pakistan and it was renamed the Central Treaty Organisation (CENTO). However on the 26th July 1956 the whole organisation was thrown into doubt when President Nasser of Egypt declared that he was going to nationalise the French and British controlled

An interesting shot of T Mk 2 WR966. This shot, probably from the early 1970s, shows the aircraft at Luqa, where it was used by No.204 Squadron from November 1971 to January 1972. This aircraft went back to the UK and was scrapped by June 1973. Note what looks like artwork on the nose?
Godfrey Mangion

Universal Suez Canal Company. Fortunately the attempt to seize the Suez Canal had been foreseen by both nations and a plan was already outlined should such an event take place. The operation, called Operation Musketeer, was put into action and announcements were made from London and Paris on the 31st October 1956. The Shackletons in the area were called upon to undertake maritime patrols in support of the operation, but once again they were used as troop transport aircraft. Operation Challenger was put into effect to uplift troops from the UK in support of Operation Musketeer. In this role five Mk 1 and 1As of No.206 Squadron uplifted the 16th Parachute Brigade from Blackbushe and delivered them to Cyprus. For the operation British aircraft had yellow and black stripes applied to the wings and fuselage. Some of the Shackletons involved were given their stripes, but not every machine was. No.38 Squadron undertook reconnaissance of Egypt from September 1956, as well as anti-submarine protection for the invasion fleet during the landings. By the 6th November the whole 'Suez Crisis' was at an end and a ceasefire was called, while a United Nations Emergency Force took over from all the ground forces in the area. The Shackleton was once again used to transport the troops, this time back home, with Nos.294 and 228 Squadron joining No.206 in the task this time around.

Oceanic Survey Work & Thermonuclear Tests

1956 was the International Geophysical Year and as part of this, No.204 Squadron took part in ocean survey work throughout the Summer.

The first use of Shackletons in true support of nuclear tests took place when four modified MR Mk 1s of No.269 Squadron were detached from their Ballykelly base to Darwin in Australia. Here the aircraft

undertook various weather data gathering sorties over the Timor Sea and Indian Ocean. Detachments to Alice Springs saw the type performing meteorological reconnaissance as well as patrols of those areas designated as being dangerous during the tests to ensure they were clear of shipping. These sorties were covered under Operation Mosaic and having completed the task they flew back to the UK via Sydney Harbour and Sharjah.

Agreement by the Australian Government had been given in the Summer of 1956 for the detonation of a limited yield thermonuclear device at the Maralinga range. Under the codename Operation Buffalo four aircraft from No.204 Squadron were detached to Pearce AFB in Western Australia. The first British thermonuclear device (Blue Danube) was dropped over the range by Valiant WZ366 on the 11th October 1956, and having completed the test the Shackletons returned to the UK in November.

As a consequence of American tests with the H-bomb the Australian Government decided to decline the British request for tests with a similar device in Australia. As a direct result of this two Shackletons of No.206 Squadron circled the world westwards as a preliminary to establishing Christmas Island as a base for forthcoming nuclear tests in the area. The H-bomb tests were covered under Operation Grapple, and after the initial survey Christmas Island was considered suitable for the tests to take place on. Two runways with hardstanding were constructed on the island and aircraft from No.206 and 240 Squadrons at Ballykelly were selected to support the trials. Their aircraft were modified at No.49 MU and the first detachment by No.240 aircraft was started at the end of February 1957. To get to the island the Shackletons flew via Lajes (Azores), Kindley Field (Bermuda), Charleston, Biggs and Travis AFB (USA), then on to Honolulu and finally Christmas Island. In April the aircraft returned after the drop of the first weapon in support of Operation Grapple 1. In January 1958 the squadron was to go back out to the island in support of the fifth H-bomb drop, Grapple Y and returned to the UK on the 3rd June. Their place had been taken on Grapple Y by No.204 Squadron, who had flown out to the island in May, and this unit also stayed for the sixth H-bomb test, Grapple Z. In July they were joined by a detachment from No.269 Squadron and once the final drop for Grapple Z had been achieved on the 11th September 1958, both detachments went home, arriving back in Ulster in October. It was during the Autumn of 1958 that a MR Mk 3 from No.206 Squadron undertook a proving test flight to Australia, although the type was not involved in the operations from this continent during the nuclear test programme as the MR Mk 1 was considered the best version to use in such trials/detachments.

The aircraft used by each squadron in support of Operation Mosaic, Buffalo and Grapple were as follows.

No.204 Squadron
MR Mk 1   VP263 and VP266
MR Mk 1A WB828, WB850, WB856 and WB857
MR Mk 2   WL739, WL747, WL748 and WL795

No.240 Squadron
MR Mk 1A WB823, WB826, WB828, WB835,
           WB856, WB859, WB860, WB861,
           WG507 and WG509

No.269 Squadron
MR Mk 1   VP265, VP289 and VP294
MR Mk 1A WB826, WB835, WB859 and WB860

Problems in Arabia

Once again the MR Mk 2s were called upon to undertake colonial policing during the later stages of 1957, when rebel activities in the Southern Arabian Peninsular flared up. No.42 Squadron sent a detachment of four MR Mk 2s to Khormaksar (Aden), arriving on the 7th January 1957. No.37 moved permanently to Khormaksar in July 1957. From here they operated against the rebel forces and in support of the ground forces, bombing and ground strafing as well as patrolling the Aden Protectorate borders. The squadron sent a detachment of two aircraft to Sharjah to assist the Sultan of Oman, however increased rebel activity meant that this detachment was soon recalled and replaced with Mk 2s from No. 48, 224 and 228 Squadrons.

By the end of 1957 No.42 Squadron had returned to the UK and its base at St. Mawgan, but in July 1958 it once again returned to Aden. Not long afterwards it moved to Sharjah where it undertook two months of bombing operations in Iraq to quell rebel uprisings. This work was taken over by No.224 Squadron off Gibraltar in 1959.

Indonesian Uprising

Britain and Malaya had signed a defence agreement on the 16th September 1957 and so when rebels started to cause problems in the nearby Sultanate of Brunei in December 1962, British military presence in Malaya was stepped up. A detachment of Shackletons was sent out to Changi by No.204 Squadron at RAF Ballykelly, arriving there on the 19th May 1964. The detachment consisted of three MR Mk 2s; WG555, WR964 and WR966. Once on site they undertook patrols to determine what the Indonesian land forces

Mk 2, WR967 is listed as being at No.5 MU when this photo was taken on the 25th October 1970, but this is obviously Malta, not RAF Kemble where that unit was based? Of interest is the 50/50 split in the upper wing colours of white and grey, used to cool the fuel cells in the wings when the type was used in a hot climate.
Godfrey Mangion

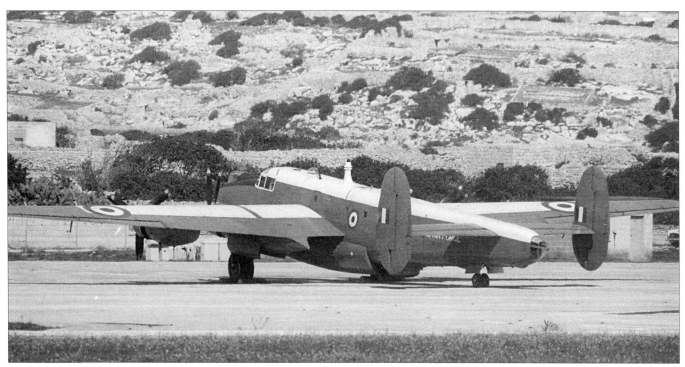

Two No.42 Squadron MR Mk 2s, complete with nose cannon, on the hard standing at Khormaksar. *Ray Deacon*

Following its Phase II update, WL756, 'C' joined No.206 Squadron at Changi in June 1962. *Barry Jones Collection*

Opposite page: MR Mk 2, WL790 seen in its final squadron service with No.205 Squadron prior to being converted to AEW Mk 2. This aircraft is still airworthy with Polar Aviation in the USA at the time of publication.

MR Mk 2, WL795 is seen in rather tropical surroundings while serving with No.205 Squadron from Changi. This aircraft was later converted to an AEW Mk 2 and is still extant on the gate at RAF St Mawgan. *Both via Martyn Chorlton*

were doing over a twelve-day period. Once this was complete, they returned to the UK and arrived back at Ballykelly on the 19th June. By August the situation was near war, so Ballykelly was again called in to support the MR Mk 2s of No.205 Squadron based at Changi. Aircraft from Nos.203 (WR965), 204 (WL739, WR964 & WR969) and 210 (WL748, WL751 and WL791) squadrons were sent. Under the codename Hawk Moth, the aircraft undertook reconnaissance missions over the Straits of Malacca. During the operations the units operated a small force off the island of Pulan Labuan, but by October 1965 the last operational sortie had been flown by an aircraft of No.203 Squadron. The Ballykelly detachment then went home and left No.205 Squadron to continue on its own. On the 13th August 1966 a peace treaty was signed and No.205 Squadron continued to operate out of Changi until they returned to the UK and disbanded in 1971, thus marking the end of Shackleton operations in the Far East.

Opposite page:

Top left: MR Mk 2, WL741, 'D' operated with No.42 Squadron in 1959-60, then No.205 Squadron at Changi and was converted to the ninth AEW.2 in April 1972.

Top right: WL738, 'M' of No.204 Squadron, made a 'good will' tour to South Africa.

Centre: WR975, 'P' of No.201 Squadron based at St. Mawgan, has yet to receive its MR Mk 3 Phase 1 modifications.

Bottom: MR Mk 3, WR973, 'U' is seen as a fully Phase 3 updated aircraft of the Kinloss Wing.
All Barry Jones Collection

This page:

An anonymous Phase 3 updated MR Mk 3, seen at the Woodford Open Day in June 1968.

Phase III modified MR Mk 3 flying at the Unison '67 Air Day, held at Boscombe Down on 7th September.

VP293 was the T Mk 4 sold to the Strathallen Museum in May 1976, but following three years of exposure to the Scottish climate, it had to be scrapped in 1979, but the front fuselage was donated to the Shackleton Association at Woodford.
All Barry Jones Collection

WB822 is seen here in T Mk 4 guise with the MOTU. It was eventually reduced to spares and used for fire practice during 1968.
via Martyn Chorlton

Chapter 6: CONTINUAL UPGRADE AND SERVICE

Phased Upgrade

The Shackleton fleet underwent a series of updates through its service career. These updates were known as the Phase Programme of Modifications. The MR Mk 1 and 1A never undertook this series of phases, as the latter was an updated version of the former anyway. The MR Mk 2 was the first to go into any form of 'Phase' modification. The prototype WB833 went to Langar on the 30th November 1954 for the installation of Phase 1.

Phase 1 of the upgrade programme saw the introduction of the ASV (Air to Surface Vessel) 21 radar system along with a Zero Reader VHF radio homer (indicated by two whip antenna on the nose), Instrument Landing System (ILS), a Mk 5 Radio Altimeter, Doppler Navigator, Search and Rescue and Homing equipment (aerials each side of nose), Mk 10 Autopilot, a Flame/Float dispenser in the beam position, an improved intercom system and all the wiring for the airborne lifeboat was removed. The Mk 2 also received the sonics plotter table that had been standard in the later mark. Mk 2s thus modified were often referred to as Mk 2C in squadron service. The first of the MR Mk 3 Phase 1 aircraft were delivered in the early Summer of 1969.

Phase 2 saw the installation of Electronic Counter Measures (ECM) equipment. This took the form of a highly visible triangular structure on top of the fuselage

that had a detachable upper section that looks a little like an electrical insulator off a telegraph pole, or some say, a lighthouse. Both UHF radio and radio homing systems were also installed with TACAN (TACtical Air Navigation) radio bearing and distance equipment, the Mk 1s Conic System associated with both active and passive sonobuoy deployment and an improved radio compass were also fitted. The HF radio antenna had to be relocated forward of the ECM tower and the first operational Mk 2 Phase 2 machines were received by the squadrons in late 1961. Again the Mk 2 airframes updated to this level also received something already present on the existing Mk 3 fleet, this time in the form of the sealed exhaust units to reduce noise levels.

Phase 3 was an involved update for the Mk 3 specifically, so I will deal with the modifications associated with this mark first, then look at what the Mk 2 got by way of a 'Phase 3' modification that differed. The Phase 3 modifications saw a considerable change in the physical form of the Mk 3, with much rebuilding. Internally the tactical table was extended and in so doing this encroached on the crew rest area so this had to be redesigned. The Radio Compass was replaced with an improved version and the Search and Rescue SARAH aerials were removed because this equipment was now replaced with the improved SABRE system. On the dorsal spine an improved VHF radio aerial was installed just forward of the ECM antenna and red anti-collision beacons were installed forward of the ECM antenna with another one underneath the rear fuselage. Navigational equipment was improved with the adoption of the GM.7 system and all the new electrical equipment meant that there was a need for more electrical power and therefore a crate of seven inverters were installed in the nose. It is odd to note that the GM.7 system worked fine in the Mk 2 Phase 3 machines, but not at all in the Mk 3 Phase 3s? In the end those in the latter type had to be replaced with the Honeywell Radar Altimeter which utilised a box-shaped aerial under the rear fuselage. Externally the wing-tip mounted tanks were increased slightly in their capacity, so they now held 253 Imp. Gal. At the same time the wiring that had originally been installed in the wings for the carriage of rockets was removed to save weight. As a result of a number of nose wheel collapses, Dunlop were approached to design a stronger nose leg unit as a replacement. The result of all this new equipment was that the all-up weight of the type rose to such a level that the power of the four Griffon 58 engines was not enough to get the aircraft off the ground at hotter, higher airfields. It was not possible for the power output of the Griffon to be increased further, so some form of additional thrust

The Viper 203 installed in the outer engine nacelles of the MR Mk 3. This shot clearly shows the doors that act as an air scoop deployed.
Hawker Siddeley

MR Mk 2, WL758, •W photographed on the 4th January 1968 while it was with No.204 Squadron. *Godfrey Mangion*

was required. It was decided that the power offered by modern jet engines would be sufficient, as they offered excellent power in relation to their weight. The decision was made to install the Viper engine produced by Bristol-Siddeley. The problems associated with installing two extra engines were many, not least of which was that the strength of the wing had to be increased to cope with the weight and thrust stresses that would be imposed by the new jet engines. The Viper 11 became the Viper 203 when installed in the Shackleton and it offered 2,700lb of thrust and weighed in at just 549lb. To reduce the need for special fuels, the Viper was tested to see if it could run on AVGAS (high-octane petrol). This showed that the engine could be fuelled in this manner, although it was restricted in the length of time it could be used at full power. The Viper could be run at a maximum setting for five minutes, although at a cruise setting it could be run for four hours. The engine was installed behind and slightly below the Griffon in the outboard nacelles. The jet exhaust was via an outlet at the rear of the revised nacelle and air was taken in to the engine via an intake in the lower edge of the nacelle. A door on hydraulic linkage folded down to act as an air scoop when the engine was running. All the controls for the Viper engines were fitted under a hinged transparent cover in the engineer's desk and the units could operate with a Time Between Overhaul of 250 hours and a operational maximum of 275 hours.

The first Mk 3 aircraft to be updated to Phase 3 standard was WR973, which moved to Woodford in July 1963. This aircraft was retained by Avro for trials with the new engines, then it was moved to Ministry of Aviation control on the 29th January 1965 to undertake flight tests. The aircraft started these trials on the 29th February and continued until the 19th May when it moved to Boscombe Down. With the A&AEE it undertook two weeks of performance testing, then it returned to Avro on the 9th June for further modifications to be incorporated (including removing all the aft sound-proofing as the aircraft was found to be tail-heavy). The aircraft went back to the A&AEE on the 28th July for the start of a full set of trials. These trials included hot weather tests in the USA, but by the 30th

September it returned to Woodford with the new Viper engine installation now cleared for service use. It was then stripped of all test and non-standard equipment and issued to the Kinloss Wing in February 1967. WR989 was also used in a series of trials at the A&AEE with the Viper. It was loaned to the unit for temperature and high humidity trials and for this it had to be skillfully flown into the stream of water being sprayed from a special tank in the bomb bay of Canberra B(I).8 (WV787) via a long ventral spray bar.

By the time WR973 was in the midst of its initial test programme, production MR Mk 3s were starting to come off the assembly lines. In the end most of the Mk 3s produced did not have the Viper installed initially and over a period of time they all went back to Avro at Woodford or Langar to have the units installed. The last Mk 3 to have the Vipers installed was XF703, which were installed at Woodford during the May 1967 to January 1968 period.

The Mk 2 airframes updated in Phase 3 were fully updated inside to Mk 3 standards, however without the added weight of the Mk 3s tricycle undercarriage they did not need the Viper engine installation, although they did need to have the power output of the Griffon Mk 58s increased and so water-methanol injection was installed. Their lower all-up weight and lack of Viper engines also meant that they did not have the tail-heavy problems of the Mk 3 Phase 3 airframes, and as a result they retained all sound-proofing aft of the crew rest area. Four in-line flare dischargers were fitted on the starboard beam and extra cabin heaters were fitted externally, just at the aft end of the radome. The first Mk 2 Phase 3 airframes started to arrive with the squadrons in the Autumn of 1966.

All Phase 3 aircraft were able to carry and drop nuclear depth charges and they all now featured Griffon Mk 58 engines, which incorporated a revision to the previous oil feed system. All the T Mk 4 training aircraft were also updated to Phase 3 standards, although they never had the revisions to the interior, nor the installation of the ECM equipment or the Viper engines.

All these major updates and revisions to the aircraft

MR Mk 2, WG555 is listed as having been operated by No.204 Squadron, but this shot clearly shows the type taking off from Luqa on Malta on the 20th August 1969.

Lovely view of MR Mk 2, WG555 on finals for landing.

MR Mk 2, WL737 was operated by No.205 Squadron at Changi when this photo was taken on the 4th August 1970.
Godfrey Mangion

meant that there was a considerable amount of retraining required for both ground and air crews. Regardless of the constraints imposed by the need to retrain the crews, in 1959 No.42 Squadron made a goodwill tour of the Caribbean, while No.224 Squadron did a similar tour of the West Indies during 1961. XF707, a Mk 3 Phase 3 machine of No.210 Squadron did a round-the-world flight during May and June 1959. This flight, called Operation Globetrotter, was to display the talents of the type to the Indian and New Zealand governments. At the SBAC show at Farnborough in 1960 a Mk 3 machine of No.201 Squadron undertook a 22 hour flight that saw it take off at the end of one days display and return the next morning for the beginning of the next. The type once again showed its ability to undertake transportation tasks, when in March 1960 it flew aid to Agadir which had been hit by an earthquake, to Belize in October 1961 after Hurricane Hatti and to Cuba in October 1961 after Hurricane Flora.

Continued Service. Policing and Potential Phase Out?

Meanwhile the Shackleton continued to see very active service in a variety of roles. No. 204 Squadron helped the security forces during the civil unrest in British Giuana during early 1962. They also did similar service in response to the Brunei Rebellion and the Indonesian Confrontation later on in that year (See Chapter 5).

During the Spring of 1965 No.201 Squadron moved up to RAF Kinloss and in July No.206 Squadron followed, with the MOTU moving down to St. Mawgan at the same time. Southern Rhodesia made a Unilateral Declaration of Independence in 1965 and renamed itself Rhodesia. What followed was to be 15 years of political wrangling and trade embargoes as well as a long drawn out guerrilla war. Strict trade sanctions were imposed against Rhodesia, and being land-locked the country depended on the port of Beira in nearby Mozambique. The United Nations heard that Rhodesia was receiving all they needed via their port at Beria, which was then being moved by land across the border into Rhodesia. The Royal Navy moved vessels in to patrol the area between Mozambique and the island of Madagascar (then called the Malagasy Republic) and the Shackleton was called upon to patrol the area for incursions. This co-operation with the Royal Navy was known as the 'Beira Patrols', although the official title was Operation Mizar. No.37 based at Khormaksar and No.38 from Hal Far were detached to the British Staging Post at Mauripur on the north-west coast of the Malagasy Republic in March 1966. The operational environment was basic in the extreme and regular flights by Bristol Britannias of Nos. 99 and 511 Squadrons were needed to keep up the flow of spares and equipment. The detachment from each unit were made up of two MR Mk 2s complete with crew and ground technicians. They were rotated on a two month cycle and operations continued until 1967 when No.38 Squadron was disbanded on the 31st March and No.37 followed suit on the 7th September. To fill the gap left by the departure of these squadrons No.42 Squadron stepped in with a detachment of MR Mk 3s. By April 1967 No.210 Squadron from Ballykelly provided a detachment of three MR Mk 3s, although now just the crews rotated every two months and the aircraft remained at Mauripur. For the next fourteen months detachment from Nos. 204 and 210 Squadrons undertook the patrols and they were then relieved by a detachment from No.205 Squadron who were based at Singapore Island, Shackleton support of Operation Mizar continued to March 1971, when WL754 of No.42 Squadron returned to the UK to mark the end of the detachments. It would not be until 1980 that a legitimate state of independence would be given and this disputed land would be renamed as Zimbabwe.

This parked aircraft is 'L' of No.205 Squadron and it is seen at Gan in the early 1960s. *Laurie Bruce*

Unidentified MR Mk 3 of No.203 Squadron powers away from the airfield at Luqa on the 4th July 1969.

MR Mk 2A, WL785 is seen here in the colours of No.42 Squadron, which was only ever based on Malta, both at Luqa and Hal Far.
via Martyn Chorlton

MR Mk 3, WR972 of the Royal Aircraft Establishment, photographed on the apron during a visit to Malta on the 13th July 1968.
All Godfrey Mangion

Changes in Policy, Home and Abroad

The 1966 Defence Review resulted in a complete revision and reduction in overseas commitment. It was anticipated that the British would be withdrawing from Aden during 1967, and there would be a general reduction in the number of forces stationed east of Cyprus. As a result of these planned changes No.224 Squadron on Gibraltar was disbanded on the 31st October 1966. In December it was decided to redistribute the maritime air forces and as a result No.38 Squadron on Malta was also disbanded. Their aircraft involved in the Mauripur detachment were recalled in

February 1967 and the whole squadron disbanded on the 31st March. No.38 Squadron was disbanded in Aden in September 1967 and this marked the complete withdrawal of maritime air force cover for the whole of the Mediterranean, as all British forces withdrew from Aden.

At this stage apart from No.205 Squadron on Gan, which offered SAR cover, there was no maritime coverage from St. Mawgan to Changhi. This was to prove a problem when unrest flared up once again in the Persian Gulf and as a result there was a need to create a detachment at Sharjah. Known as the Sharjah

Maritime Detachment this unit was controlled by the Kinloss Wing and surveillance, anti-smuggling and anti-gun running patrols were undertaken over the Persian Gulf. The detachment also used machines from St. Mawgan and Ballykelly, although the Kinloss Wing remained the main unit involved in this operation. The need for a permanent unit at Sharjah was met when No.210 Squadron reformed there on the 1st November 1970. This unit was equipped with five T Mk 2 aircraft that had previously been used by the MOTU and had been updated to Mk 2 Phase 3 standard at Kemble. These machines were WG554 (V), WL787 (W), WL739 (X), WG558 (Y) and WR967 (Z). The build-up of Russian naval equipment in the Mediterranean saw four aircraft from No.42 Squadron being sent to Malta in October 1968. They remained here until 30th January 1969 when they were relieved by No.203 Squadron who took up permanent residence on the island. The latter stages of 1969 also marked the end of Coastal Command and a Disbandment Parade was held at St. Mawgan on the 27th November. The flypast included three Shackletons from each of the UK bases (St. Mawgan, Ballykelly and Kinloss) and a Nimrod, the types replacement, brought up the rear. This was followed the next day with the reforming of the force as No.18 (Maritime) Group, Strike Command. Strike Command took over the maritime role with a force that was in a period of transition. The first Nimrod had only just been delivered to St. Mawgan and the three Shackleton squadrons at Kinloss (Nos.120, 201 & 206) converted over to the type during 1970. The final Shackleton training course took place with the MOTU on the 28th July 1970 and on the 1st August the MOTU was renamed No.230 OCU to undertake crew conversion training for the Nimrod. The last MOTU Shackleton, XF703 (42/J), was held back for the Battle of Britain display at St. Mawgan in September 1971 and this aircraft was then flown into store at RAF Henlow for the proposed RAF Museum. Unfortunately this machine was later scrapped by them.

Because the Nimrod had been deemed too expensive to use as an SAR aircraft, the Mk 2 Phase 3s of No.204 Squadron were assigned the task. This unit disbanded at Ballykelly on the 1st April 1971 and reformed at Honington the same day. Here the unit comprised two flights, the first of six aircraft for SAR duties and the other with two aircraft for the Mauripur Detachment Support Unit. It is pleasing to note that Ballykelly, a base long associated with the Shackleton, was handed over to the Army on the 2nd June 1971 and was renamed as Shackleton Barracks.

In the Far East the 1966 Defence Review plans were now in full force and as a result No.205 Squadron at Changhi disbanded on the 31st October 1971 and nearly all of their aircraft were flown back to the UK. Four machines remained at Changi to offer SAR cover and these were operated as a detachment of No.204 Squadron. This was followed on the 15th November by the disbandment of No.210 Squadron at Sharjah and their three remaining aircraft (WG554 and WG558 already having returned to the UK) were flown back to RAF St Athan. Land disputes with Malta saw No.203 Squadron moving from the island to Sigonella in Sicily in December, having re-equipped with the Nimrod whilst on Malta in October. Their last two remaining Shackleton Mk 3 aircraft were flown back to MUs in the UK, so by the end of 1971 there were only twelve Mk 2 Phase 3s with No.204 Squadron, six at Honington, two with the detachment at Mauripur and four still on detachment at Changhi. Several were also held inactive at MUs in the UK. By February 1972 the United Nation blockade on the port of Beira was lifted and when SAR duties were transferred to the Nimrod squadron, No.204 Squadron disbanded at Honington on the 28th April. No.203 Squadron was able to move back to Malta though, arriving at Luqa on the 23rd April 1972, as the previous land disputes had been resolved.

MR Mk 3, WR974, photographed on the 16th March 1968.
Godfrey Mangion

MR Mk 3, 'F' of No.42 Squadron taking off from its base on Malta, 6th January 1969.

MR Mk 3, WK986, 'G' of No.203 Squadron coming in to land on Malta on the 15th February 1969.

MR Mk 3, XF708, 'C' of No.203 Squadron at Luqa while taxying out on the 16th November 1971. *All Godfrey Mangion*

MR Mk 3, 'C' of No.203 Squadron on finals to Luqa on the 21st August 1968. Note the intakes for the Vipers are open.
Godfrey Mangion

MR Mk 3, 'E' of No.203 Squadron, at Luqa in November 1971.
Godfrey Mangion

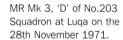

MR Mk 3, 'D' of No.203 Squadron at Luqa on the 28th November 1971.

MR Mk 3, WR974 in company with another No.203 Squadron example at Luqa on the 16th September 1969.

MR Mk 2, WL747 of No.204 Squadron comes in to land at Luqa on the 8th September 1970. This aircraft returned to the UK in December that year and was converted to a AEW Mk 2.

MR Mk 2, WL748 tucks up its undercarriage when taking off from Luqa on the 5th June 1969.
Godfrey Mangion

An unidentified MR Mk 3, probably of No.203 Squadron taxies in at Luqa on the 7th September 1969.

MR Mk 2, WL755 seen with No.204 Squadron on the 19th March 1970. This aircraft became part of the Majunda Detachment Support Unit from July 1971, before returning to No.204 Squadron. It finally ended its days on the fire dump at Catterick in December 1977.

MR Mk 2, WL756 from No.204 Squadron at Luqa, seen here on the 21st February 1970. It was converted to AEW Mk 2 in April 1971 and served with No.8 Squadron until burnt at St. Mawgan in 1998. *All Godfrey Mangion*

Top: MR Mk 2, WL756 from No.204 Squadron at Luqa, seen here on the 29th August 1970. Note the painted out area on the fuselage side (possibly a unit marking from its time at Ballykelly) and the lack of the cumbersome aerial antenna.

Above: MR Mk 2, WL756 from No.204 Squadron seen at Luqa on the 29th August 1970.

Below: Ex-No.38 Squadron and Ballykelly Wing MR Mk 2 WL758 seen during its time with No.204 Squadron at Luqa. Photographed here on the 4th August 1970, note the painted out crest on the nose, probably a relic of its time with either of the previously mentioned units.
All Godfrey Mangion

No.228 Squadron's MR Mk 2, WR957, 'U' photographed in the 1950s, prior to receiving its Phase 1 update.

MR Mk 2, WR961, 'S' was fully updated to Phase 2 standard when seen with No.38 Squadron at Luqa in 1964.

Daily twenty-two hour MR Mk 3 patrols were a part of the RAF's participation at the 1960 SBAC Display and Phase 1 updated XF711, 'L' is seen returning to Farnborough. *All Barry Jones Collection*

Chapter 7: **THE PROJECTS, THE 'INTERIM SOLUTION' AND THE FOLLY**

The Mk 4 & 5 Projects

Development of the engines in the Shackleton to increase range was not likely, as Rolls-Royce in the post-war period was little interested in further development of their existing piston aero engines, with all future development centred around jet and rocket engines. Prior to WWII, the engine manufacturer D. Napier & Sons Ltd had obtained a licence agreement with the German engine manufacturer Junkers Motorenbau, especially for their diesel aero engines. Initial trials used the Jumo 204A, which was tested in Hawker Horsley II J8620, although official interest was not sufficient to get any orders or development contracts. After WWII ended Napier resumed their interest in the compound piston/gas turbine engine and developed the E.125 Nomad NNm1. This new compound engine was being heralded as 'the' engine for endurance and economic performance and as a result Avro proposed to use the type in the Shackleton and thus the MR Mk 4 project was created.

The project was designed around a MR Mk 1 airframe and would be done in two stages; first with just two Nomad engines in place of the outer Griffons (Avro Type 717) and finally as the true MR Mk 4 with Nomads in place of all four Griffon engines (Avro Type 719). The Nomad was an extremely complex design, probably conceived before its time. It was a twelve-cylinder, horizontally-opposed, liquid-cooled, compression-ignition diesel inline fitted with an eleven-stage, axial-flow compressor driven by the exhaust across a three-stage gas turbine. This compressor stage drove the front section of the six-blade, contra-rotating propeller. The type was first tested in the nose of a Lincoln B Mk 2 (SX793) and shown to the public at the SBAC in 1951. The 3,200shp engine

could propel the lightly-fuelled Lincoln all on its own and SX793 flew several passes with just the Nomad turning. The engine type in its initial form was too complex and bulky, so Napier revised it and offered it as the E.145 Nomad NNm.2. This unit coupled the diesel units and the compressor to drive a single four-blade Rotol or De Havilland 13ft diameter propeller via a variable gear unit. This engine was run for the first time in December 1952 and a trial installation of it was agreed by the Ministry of Supply. An initial order of six units was also authorised. Trials were undertaken at Luton Airport by Napier in 1953, where the second Shackleton prototype VW131 flew in to be fitted with the engine. Initially just a mock-up of the Nomad was used to sort out problems with installation and other associated ancillary equipment, but by April 1952 vibration tests were being undertaken with a real unit in place. Unfortunately just as the installation was to take to the air, with two E.145s in the outer engine locations on VW131, the whole project was cancelled. This was all brought about by Treasury demands for severe cuts in British aircraft industry spending. VW131 remained at the test shed at Luton for nearly two years before this important prototype was finally dismantled. The fuselage went to Avro's Bracebridge Heath facility, where it was used for Shackleton ditching trials and finally tested to destruction and SOC on the 10th May 1962. The MR Mk 5 is said by many to be a spurious designation that relates to a design that never existed, while others state that Avro Type 719 related to the Shackleton MR Mk 5 complete with its four Nomad NNm.6 engines?

Other powerplants were also considered for the Shackleton. One of these used four Wright Cyclone R3350-85 engines. This machine was similar to the

A lovely drawing from Napier of their Nomad.

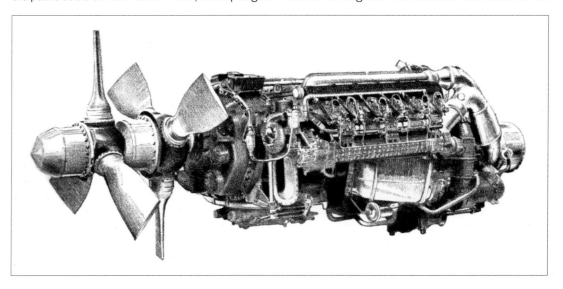

THE PROJECTS, THE 'INTERIM SOLUTION' AND THE FOLLY

MR Mk 3, but had a large single fin and rudder assembly with an extending ventral fin and dihedral tailplane. The radar was to be fitted in a ventral radome similar to the AEW Mk 2, although it was to be positioned further aft than that one, however this project went no further than the drawing board.

AEW Mk 2

By the end of 1971 all of the MR Mk 3s were out of service, simply because the stresses exerted on the airframe by the Viper jet engines were such that the entire airframe soon exceeded its fatigue life. The Shackleton was not to die though, because in the late 1960s it had become apparent that there was a real need for an airborne early warning platform and the Mk 2s were chosen for the task. In 1967 it had been agreed that a number of MR Mk 2s should be converted to the AEW role. A check of all existing stocks of MR Mk 2s was undertaken, to see if there were enough with low flying hours, and in the end twelve MR Mk 2 Phase 3s were chosen for conversion.

The conversion process started with the aircraft moving to No.5 MU Kemble with each airframe flown in from their respective squadrons. While at No.5 MU each airframe was inspected and any repairs and/or overhaul took place. The airframes were then put in store until they were called forward by the Hawker-Siddeley works at Bitteswell. The first AEW Mk 2, WL745, flew for the first time at Woodford on the 30th September 1971.

No.8 Squadron was chosen to operate the AEW Mk 2. It had disbanded at Muharraq, so it was reformed on the 1st January 1972 as part of No.11 Group, Strike Command. It is strange to think that this famous fighter squadron would operate a type with a lineage back to the Manchester and Lancaster. It is even stranger to think that as a fighter unit the squadron operated their Venoms and Hunters alongside the Shackletons of No.37 Squadron in Aden! Initially the unit also operated a couple of Mk 2s in the trainer role, the first being WL787 which arrived on the 1st November 1971. The unit was based at Lossiemouth, although when this base was having the main runway resurfaced and strengthened, they also operated from nearby Kinloss. The first AEW Mk 2 to come to the unit was WL747 on the 11th April 1972 and it was followed by WL756 and by the end of the year the unit had eight AEW Mk 2s on charge. The final AEW Mk 2 taken on charge by the unit was WL745 which did not arrive until the 17th September 1973, having been involved in Ministry of Supply and A&AEE trials since March 1970. With work finished at Lossiemouth No.8 Squadron officially took up residence there on the 14th August 1973.

The operation of the Shackleton as an AEW platform was always considered an interim measure, but apart from the type being brought into service to fill the void before the Nimrod AEW arrived, the reason such an old type was used was simply that it was cheaper to

convert twelve existing (out of service) types to the role than to buy purpose-built ones from elsewhere. The Shackleton was only initially going to be used to offer AEW cover for the fleet, as the Royal Navy had lost its AEW cover with the disposal of its aircraft carriers and their associated fleet of AEW Gannets. However in the end the roles for the AEW Mk 2 consisted of the initially envisaged AEW for the fleet plus AEW within the UK air defence region, control of air-defence aircraft, surface forces observation via electronic surveillance measures (ESM), direction and control of strike/attack aircraft and even a limited capacity in the search and rescue role, as all of them carry Lindholme rescue gear. When controlling air-defence aircraft the Shackleton works in conjunction with various Master Radar Stations in the UK to guide them throughout such interception. One of the main problems initially with the type was 'clutter' (signal returns) caused by the sea and its effect on the radar systems within the aircraft. An Airborne Moving Target Indicator (AMTI) was fitted after successful trials at RRE Pershore to cut down the amount of such returns from the sea. These returns clutter the radar scope and depending on atmospheric conditions etc can effectively reduce the radars range considerably.

The AP/ANS 20 radar fitted in the Shackleton AEW Mk 2 was based on the system developed in

A view form the side of the revised nose of Lincoln SX973 with the Nomad installed. *via J Grant*

This shot shows a Napier Nomad installed in the port outer nacelle of Shackleton VW131 in April 1954. *Napier*

AEW Mk 2, WL754 of
No.8 Squadron
photographed at Luqa
during a visit there on the
14th February 1973.

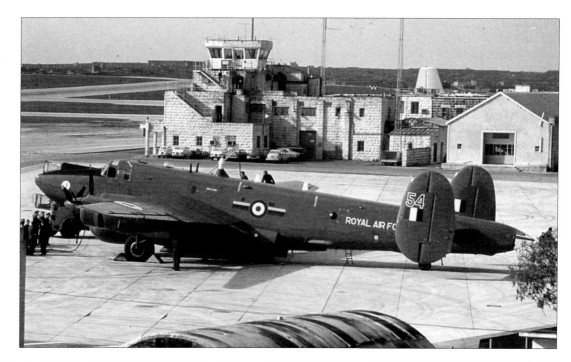

AEW Mk 2, WL754 in the
company of WL793, both
of No.8 Squadron,
photographed on the 2nd
May 1973.
Both Godfrey Mangion

America during the 1940s. The ANS 20-based system fitted in the Shackleton is the F series, which was a much evolved version from the original. The power required by this system was 2mW and this was generated by two a.c. generators fitted to numbers one and two engines. The system had an effective range of about 200 miles. The scanner scope image was ground-stabilised from the aircraft Doppler, and was also north-stabilised so that north was at all times at the top of the display. AEW operations even in the early 1970s were a fairly heavy electronic counter-measure environment, so the AEW Mk 2 retained the maritime Shackleton ECM Orange Harvest system to counter this. Other electronic systems initially installed included the APX 7 IFF (identification friend or foe) and both passive and active SIF (selective identification facility). Radio equipment included two PTR 175 V/UHF, two R52 UHF and two Collins 618T single-side-band HF sets. The AEW retained its large bomb bay,

although this was usually filled with smoke and flame flares and Lindholme gear. The most noticeable external change with the AEW version was the addition of the large ventral radome that houses the radar. This new radome replaced the older 'dustbin' unit of the MR Mk 2 and it was relocated at the front edge of the bomb bay. The 20mm cannon were also deleted. The AEW Mk 2 had a crew of nine, consisting of pilot, co-pilot, radio navigator, 'navigating' navigator, engineer and four radar operators. The latter come in three groups: the basic AEW operator who is trained but requires supervision; the more experienced operator who can work without supervision; and the Tactical Co-ordinator (Taco) who is the most experienced operator and is also responsible for managing the other operators while in a tactical situation and generally managing the aircraft system in the most effective manner throughout a mission. When the squadron was established at Kinloss in early 1972 there was also an OCU on site, which used a couple of Mk 2s as trainers. Their first task was to combine a certain number of experienced Shackleton crews with Gannet AEW observers to make up enough crews for the new aircraft. Initially the Royal Navy loaned twenty-six AEW operator/controllers to the RAF for this purpose, although over time these were gradually replaced with RAF operators. Initially seven twenty-four day short conversion courses were used to get the crews up to speed, but this was replaced by a standard 15-week course and by early 1973 there were nine operational crews and eleven operational aircraft.

In 1981 the then Conservative Government decided to implement a series of defence cuts and No.8 Squadron saw their fleet cut in half. The six highest flying hour airframes were taken off their inventory and these were WL741, WL745, WL754, WL795, WR860 and WR963. Of these, WL741 and WL745 were burnt at Manston and Catterick respectively. Having weathered such a heavy cut the unit soldiered on for another ten years in the role, but had to suffer a tragic loss less than one year before the type finally left service.

WR965, piloted by the squadron's CO, Sqn Ldr Stephen Roucoroni, crashed into a hillside near Northron in the Outer Hebrides on the 30th April 1990. All crew members on board were killed, it was a bitter pill to swallow considering the excellent safety record the Shackleton had throughout its life.

Nimrod AEW Mk 3 and the Boeing AWACS

The whole subject of Airborne Early Warning systems had been discussed by all nations since WWII, and in America the E-2A Hawkeye with its dorsal-mounted radar dish and the Boeing E-3 AWACS based on the Boeing 707 were all good examples of the concept. Although by the early 1970s NATO had decided to opt for the E-3, the British Government decided to go it alone and develop a purpose-built AEW based on the Nimrod. The project involved British Aerospace and GEC Marconi and the types radar scanners were fitted in two big bulges in the nose and tail. The radars used were the GEC Marconi Argus System with two mechanical scanners which gave 360 degree coverage of air and surface targets. Redundant Nimrod

AEW Mk 2, WL741 of No.8 Squadron photographed on the 20th February 1974.

AEW Mk 2, WL756 of No.8 Squadron seen in the company of another squadron machine and Nimrod MR Mk 1s in February 1977.
Both Godfrey Mangion

This is the 'Think British' badge produced to promote the AEW Nimrod in the 1980s.
Author's Collection

Two No.8 Squadron AEW Mk 2s (including WR960) line-up alongside a Handley Page Hastings while visiting Malta.
Godfrey Mangion

MR Mk 1s were used for the project, converted structurally by British Aerospace. The aircraft with its bulbous nose and tail flew successfully but during the Joint Trials Unit (JTU) work-up serious problems were encountered with the radar system. Software difficulties combined with a lack of co-ordination of the two scanners and their data feeds were paramount. The GEC 4080M on-board computer was seen to be the main problem, and it was estimated that a 300% increase in its capacity would only allow it to do what was already being asked of it, and would give no potential for future development! By 1985 the Government had changed from the Labour one that had originally decided to produce the Nimrod AEW and much play was made of the delays and escalating costs. By this time GEC were still saying that the system problems could be ironed out and although the twelve Nimrod airframes had been delivered by British Aerospace the actual costs were out of control related to the development work being undertaken by various sub-contractors with regard to the electronic systems. By the end of 1984 the whole project had cost £816 million and so the Ministry of Defence set in place a new fixed-term contract that saw a set fee and delivery to operational standard 'some time in 1987'. By the end of 1986 the three development aircraft (P1, P3 & P4) were all involved in trials trying to perfect the avionics with the JTU at Waddington. Despite great pressure from political, military and business quarters, the JTU was able to clearly show that the system did not work and an emergency meeting was held in the House of Commons in January 1987. The outcome of this meeting was a 169 majority for the Government and they therefore cancelled the whole Nimrod AEW.3 project. The remaining Nimrod AEW airframes were ultimately scrapped, as it was considered that they were too extensively modified to be returned to their original specification. In all over £900 million had been spent on the Nimrod AEW project. Many state this was for purely political reasons and it is true to say that the whole subject was seen as a needless waste for the tax payer, but at the end of the day the type was crippled from the outset by requirements it tried to meet but which were never going to be realistic. In 1988 the British Government accepted a Boeing proposal for the supply of seven E-3D Sentry (AWACS) aircraft at a cost of £860 million pounds.

GEC Marconi continued to develop the Argus System and the Chinese Air Force has signed agreements with them for a development of it to provide an AEW capability for their air force and it is most likely that this system will be installed in the Ilyushin Il-76 'Candid' for this role.

The End of the Line

The first Boeing Sentry AEW Mk 1 (ZH102), as the type was known, arrived at Waddington in early November 1990. Elements of No.8 Squadron had moved down from Lossiemouth and with the official handing over of the squadron colours from Lossiemouth to Waddington, the operational career of the AEW Mk 2 and the Shackleton as a whole came to an end.

It had been 30 years since the Shackleton was described as 'worn out and obsolete' by the press and 19 years after being established as a 'temporary expedient' in the AEW role, but at long last the faithful old Shackleton went to her long overdue rest.

Today, of the 180 built only very few are left. Below is a list of those that have been preserved.

Survivors: United Kingdom

• T Mk 4, VP293, nose section only, now on display at the Newark Air Museum.
• T Mk 4, WG511, nose section only on display at the Flambards Village Theme Park, Helston, Cornwall.
• AEW Mk 2, WL756, nose section only, St. Austell, Cornwall.
• AEW Mk 2, WL795 gate guard at RAF St. Mawgan, Cornwall.
• AEW Mk 2, WL798, forward fuselage, privately owned, Elgin.
• AEW Mk 2, WR960 static display at the Museum of Science and Industry in Manchester, Air and Space Gallery, Manchester.
• AEW Mk 2, WR963 owned by Air Atlantique and based at Coventry Airport. Radome removed and repainted as an MR Mk 2.
• MR Mk 3 Phase 3, WR971, fuselage only on display at the Fenland and West Norfolk Aviation Museum,

Walton Highway, near Wisbech, Cambridgeshire.
• MR Mk 3 Phase 3, WR974 static display at the Gatwick Aviation Museum, Charlwood, Surrey.
• MR Mk 3 Phase 3, WR977 on static display at the Newark Air Museum, Nottinghamshire.
• MR Mk 3 Phase 3, WR982 static display at the Gatwick Aviation Museum, Charlwood, Surrey.
• MR Mk 3 Phase 3, WR985 static display with the Jet Aviation Preservation Group, Long Marston, Warwickshire.
• MR Mk 3 Phase 3, XF708, static display at the Imperial War Museum, Duxford, Cambridgeshire.

Cyprus

• AEW Mk 2, WL747 owned by Savvas Constantinides. Stored at Paphos.
• AEW Mk 2, WL757 owned by Savvas Constantinides. Stored at Paphos.

South Africa

• MR Mk 3, S/No.1717 dismantled on farm Stangar, Kwazulu, Natal.
• MR Mk 3, S/No.1720 (M), SAAF Museum, Ysterplaat AFB.
• MR Mk 3, S/No.1721 on display at the SAAF Museum, Swartkop, Pretoria.
• MR Mk 3, S/No.1722, airworthy with the SAAF Museum, based at Ysterplaat AFB.
• MR Mk 3, S/No.1723, this is the example that is at a garage on the freeway, Johannesburg painted like a huge Coca-Cola bottle!

United States

AEW Mk 2, WL790 (N790WL), owned by Air Atlantique and operated by Amjet, Blaine, Minnesota.

AEW MK 2 prototype WL745 touches down with puffs of smoke from the main wheels.
Godfrey Mangion

Top: An early shot of the AEW Mk 2, WL745, when it still retained the grey and white scheme.

Above: AEW Mk 2 prototype WL745 seen in its early colour scheme during tropical trials on Malta.

Below: AEW Mk 2, WL754 of No.8 Squadron shares the ramp with another example and a Handley Page Hastings on the 2nd July 1973.
All Godfrey Mangion

Lovely shot of No.8 Squadron AEW Mk 2, WL754 coming in to land on the 10th May 1973. Note the blanked off space where the ventral radome was on the MR Mk 2.

AEW Mk 2, WL790 of No.8 Squadron seen here landing at Luqa on the 22nd February 1977.

AEW Mk 2, WL790 of No.8 Squadron seen coming in to land at Luqa on the 22nd February 1977.
Both Godfrey Mangion

AEW Mk 2s WL754 and WL795, both of No.8 Squadron, seen here in the company of a Handley Page Hastings, at Luqa on the 2nd July 1973. The former ended its days at RAF Valley, while the latter still exists at St. Mawgan. *Godfrey Mangion*

AEW Mk 2, WR960 showing the various modifications made to the airframe for its role as an AEW platform as well as clearly showing the final camouflage and marking scheme applied to the Shackleton. This aircraft is still extant at the Museum of Science and Industry in Manchester. *via Martyn Chorlton*

AEW Mk 2, WR965, of No.8 Squadron seen taxying out at Luqa on the 2nd July 1974. *Godfrey Mangion*

Overall view of No.8 Squadron AEW Mk 2 WL745 on Malta. Note the coloured bars either side of the fuselage roundel, a kick-back to No.8 Squadron's fighter days.

AEW Mk 2, WL754 of No.8 Squadron was called 'Paul'.

No.8 Squadron AEW Mk 2, WL754 in the company of another No.8 Squadron aircraft on the hardstanding at Luqa, Malta.

AEW Mk 2, WL756 was called 'Mr Rusty' while in service with No.8 Squadron.
All R J Caruana

Chapter 8: **SHACKLETON COLOURS**

The Shackleton, like all frontline aircraft, has undergone a number of camouflage and marking changes over the years, so what follows is a brief description of the various schemes applied to the type in both UK and South African service.

Coastal Command

A directive issued on the 10th August 1941 had decreed that all Coastal Command aircraft would be painted white overall with all the upper surfaces visible when viewed from directly above finished in a camouflage pattern of Dark Slate Grey and Extra Dark Sea Grey. In the immediate post-war period the upper colours were gradually replaced with Medium Sea Grey and it is in this scheme that the prototype Shackletons appeared. Serial numbers were painted in Light Slate Grey, aft of the roundel on the rear fuselage and roundels were carried on either side of the rear fuselage as well as on the upper and lower outer wing panels. The serial number was repeated in Light Slate Grey initially under each wing, inboard of the roundel, but this was later changed to black. Local variations occurred and Joint Anti-Submarine School (JASS) machines were seen with black bands chord-wise across the outer wing panels, while some units applied a broad black band, broken by the roundel, around the rear fuselage.

In 1955 the rules changed once again and all maritime aircraft were to be finished in Dark Sea Grey overall. The serial number and unit codes were to be applied in red with white outlines. Some machines had the radome painted in Medium Sea Grey, but this was by no way standard across the entire Shackleton fleet.

In 1959 this scheme was changed yet again and the upper fuselage surface was now to be painted white to reduce the interior temperature. All MR Mk 1As and T Mk 4s saw out their service lives in this scheme. VP293, the T Mk 4 operated by RAE Farnborough was different though, as it wore the above overall scheme but also featured the nose (except the radome, which remained grey), fins and rudders, rear turret fairing and wing leading edge panels in day-glo red (later painted with plain orange, as the day-glo faded too quickly), plus the engine cowlings and panel forward of the cockpit in black. In the early days the leading edge day-glo area did not extend to the panels between the wing and upper engine cowling, as these remained in grey, but at some stage in its life these areas were also painted day-glo, as by the time the aircraft retired to Strathallen, these areas were orange. The propeller blades had alternate black and white bands, although these only applied to the outer set of the outer engines and the inner set of the inboard (port) engines; for some reason the inboard starboard propellers were not marked in any manner other than the usual red, white, red tips? The spinners also had a complex colour coded system with the front section of the inner engines and inner sections of the outer engines in day-glo red (orange) and the inner sections of the inner (port) engine in white, while the outer sections of the outer engines were painted 50/50 white and black! The red fin end plates also had the fin flashes (both inner and outer faces) outlined in white and no roundels were carried under the outer wing panels. The aircraft carried the serial in day-glo red (orange)

MR Mk 1A, WB822 of No.236 OCU in the earliest scheme used by the Shackleton. This aircraft was later converted to T Mk 4 before returning to the MOTU.
via Martyn Chorlton

low down on the rear fuselage side and under each outer wing panel in black

Unit badges etc were not common on the MR Mk 1, 1A or T Mk 4 aircraft, although once the darker grey scheme was applied Nos. 120, 240 and 269 Squadrons and some of the MOTU T Mk 4s did carry unit badges on the nose.

The MR Mk 2 production batch were all finished in the white and Medium Sea Grey scheme. Roundels were only applied to the fuselage and upper wing surfaces and the serial number was applied in black aft of the fuselage roundel and under each outer lower wing panel (read from the front on the starboard side and from the rear on the port side). Adoption of a new identification system came into being in the early 1950s that resulted in a single letter for the unit code and an individual letter for each aircraft. The former was placed to the rear of the fuselage roundel with the latter forward of it. Photographic evidence also shows that a slightly different system was used with the squadron letter remaining well aft of the fuselage roundel and the aircraft letter being placed either side of the nose. In the latter case, if the unit had too many aircraft for letters in the alphabet, then it was common to add another number alongside the letter in a smaller size character e.g. Z1.

As with the previous marks the scheme changed to the overall Dark Sea Grey one in 1955 and the squadron identification code system was dropped and replaced with a simple one that just used the numbers of the squadron e.g. 37, 224, 228 etc. These were applied in white outlined red characters on either side of the rear fuselage and the roundels previously seen in this area were dispensed with (the roundels on the upper wings were retained though). The individual aircraft letter was applied in the same manner either side of the nose and the serial number, also in white outlined red, was applied under each outer wing panel. In 1956 the overall size of the squadron number was reduced to allow the 'Royal Air Force' legend to be

applied aft of it on either side of the rear fuselage, with the serial number in small red characters below it. The exception to the rule here was No.38 Squadron at Luqa, as they kept the squadron number large and added the 'Royal Air Force' in small letters on the fins end plates. At the same time propeller spinners were painted in squadron and flight colours.

One of the most adorned aircraft was T Mk 2 WR966, which was initially used by No.205 Squadron, then was moved to the No.205 detachment at Tengah. As this unit was a Far East Air Force detachment the aircraft was painted with a mass of markings before it came back to the UK. The No.205 Squadron crest was applied on the fin end plates and the legend 'White Knuckle Airlines' applied on the fuselage sides just below the demarcation of the white and dark grey. The individual aircraft code T was amended to GT (as in Grand Tourismo!) and '146,800cc' was written alongside it in smaller characters. Painted in this manner the aircraft left Tengah in early January 1973 and arrived at No.32 MU St. Athan on the 10th January. It remained in store at St. Athan until the 22nd June 1973, when it was SOC and scrapped. A pity as it would have made a very eye-catching exhibit.

The MR Mk 3 entered service in the overall Dark Sea Grey scheme. In 1960 the white upper fuselage decking was applied plus the individual aircraft number on the nose and squadron number on the rear fuselage. Squadron embellishments came in the form of the Union Jack on the nose or tail end plates and No.201 Squadron added the squadron crest above the fin flash. Both No.201 and 203 Squadron, plus some of the aircraft of No.206 Squadron, had the tip tanks painted red. No.206 Squadron followed No.201 but had their crest underneath the fin flash. With the adoption of the Central Servicing policy, the use of squadron numbers was dropped and most of the Scottish units that used the system replaced the individual squadron-linked number with just the 'Royal Air Force' legend in white on the rear fuselage.

The RAE-operated T Mk 4 VP293 is seen in flight. Its colourful overall scheme is obvious. *RAE*

The MR Mk 3 retained by the Controller (Air) and operated by the RAE (WR972) had a wide variety of schemes applied to it throughout its life. Initially this aircraft retained the standard Coastal Command 'grey' scheme, but shortly after arriving with the RAE on the 13th April 1961 it was repainted silver overall with day-glo red panels applied to the nose, wing tips and tip tanks, fin end plates and spinners. The undersurfaces were finished in broad bands of yellow and black, as was fitting for a test airframe. During March 1968 this scheme was changed to conform more closely with the then standard RAE scheme. The upper fuselage and wings were painted white and the lower fuselage and uppersurfaces of the wings were painted light grey. Later a dark blue cheat line was applied at the demarcation of the upper white and lower grey and it extended back to the tail, with a break for the serial number which was also dark blue, and forward to the nose. The blue also covered the tail end plates with the fin flashes outlined in white, the tip tanks, engines nacelles and across the top of the wing. In this area it ran chord-wise across the wing in line with the engines, very much like exhaust staining. Also at this time, in dark blue, the legend 'Royal Aircraft Establishment' was applied just above the cheat line on either side of the rear fuselage.

All of the twelve AEW Mk 2s converted were painted at No.5 MU Kemble in an overall Dark Sea Grey gloss colour with a matt black anti-dazzle panel forward of the cockpit. The squadron badge comprising a round white surround on which is mounted a Jambiya curved dagger, was applied high up on either side of the nose. The 'Royal Air Force' legend was applied in white on the rear fuselage, with the serial number in red just below it. The last two digits of the serial were repeated in white-outlined red characters above the fin flash on the end plates. As No.8 Squadron had originally been a fighter squadron, this was identified by the adoption of yellow/blue/red stripes applied either side of the roundel on the fuselage. In 1987 a large white-outlined red number 8 was applied ahead of the 'Royal Air Force' legend on either side of the fuselage. Propeller blades on the AEW Mk 2 were tipped red, white, red and the aircraft featured various emergency and 'cut here' markings on the airframe, usually in yellow.

South Africa

The first three aircraft that were delivered to South Africa were painted Extra Dark Sea Grey on the upper surfaces, with a close approximation of PR Blue on the fuselage sides and lower surfaces. Each aircraft had the serial number in black characters on either side of the rear fuselage, aft of the roundel and an individual aircraft letter in yellow was applied to either side of the mid-fuselage, just forward of the roundel. The aircraft serial number was also duplicated under the outboard wing panels, just inboard of the roundels and these could be read from behind on the starboard side, and from in front on the port side. The aircraft carried the 'new' Springbok roundel of the SAAF which had been introduced in 1951 and the No.35 Squadron emblem of a Pelican on a stylised map of South Africa was applied to each side of the nose.

Only the first three machines delivered carried the Springbok-style roundel. The remaining five machines all featured the revised roundel known to all as the

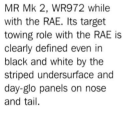

MR Mk 2, WR972 while with the RAE. Its target towing role with the RAE is clearly defined even in black and white by the striped undersurface and day-glo panels on nose and tail.

'Castle' roundel. This featured a stylised ground plan of the Cape of Good Hope in dark blue with a white surround, onto which was centred a gold-coloured leaping Springbok. This leaping animal ran right to left on the port side and left to right on the starboard. The SAAF national insignia changed again in 1981, with the replacement of the Springbok with a gold eagle motif, although this was never applied to the Shackleton fleet in SAAF service.

The overall scheme for the type changed not soon after arrival, with the adoption of the white upper decking to reduce the interior temperature. Later the approximate PR Blue colour was extended up over all the flying surfaces and tail and rudder assemblies. At the same time all the undersurfaces were repainted in Dark Sea Grey and all the propeller spinners were painted red. At this time it also seems that the last two digits of the serial number were added in black characters on the extreme tip of the nose, between the cannon barrels and a matt black anti-dazzle panel was added on the upper nose section forward of the cockpit.

Above: AEW Mk 2, WL754 of No.8 Squadron photographed coming in to land on Malta on the 4th May 1973. *Godfrey Mangion*

Top: WG530, the first production MR Mk 2 from No.205 Squadron based at Changi.

MR Mk 1 & 1A

Avro Shackleton MR Mk 1, WB854/B-S, No.224 Squadron RAF Gibraltar, early 1950s.
Medium Sea Grey/White scheme with Light Slate Grey codes and serial on fuselage sides;
serial under wings in black. Note non-standard fin and rudders in Medium Sea Grey.

Avro Shackleton MR Mk 1A, WG511/A-I, No.42 Squadron, summer 1952.
Medium Sea Grey/White finish with Light Slate Grey codes and serial on fuselage sides;
serial under wings in black.

Avro Shackleton MR Mk 1A, WB851/G-V, Joint Anti-Submarine School, RAF Ballykelly.
Medium Sea Grey/White scheme with Light Slate Grey codes and serial on fuselage sides;
serial under wings in black. Red front of port outer spinner.

Avro Shackleton MR Mk 1, VP262/A-D, No.120 Squadron, RAF Aldergrove, 1953.
Medium Sea Grey/White scheme with Light Slate Grey codes and serial on fuselage sides;
black serial under wings. Black spinners; 'D' repeated on nose.

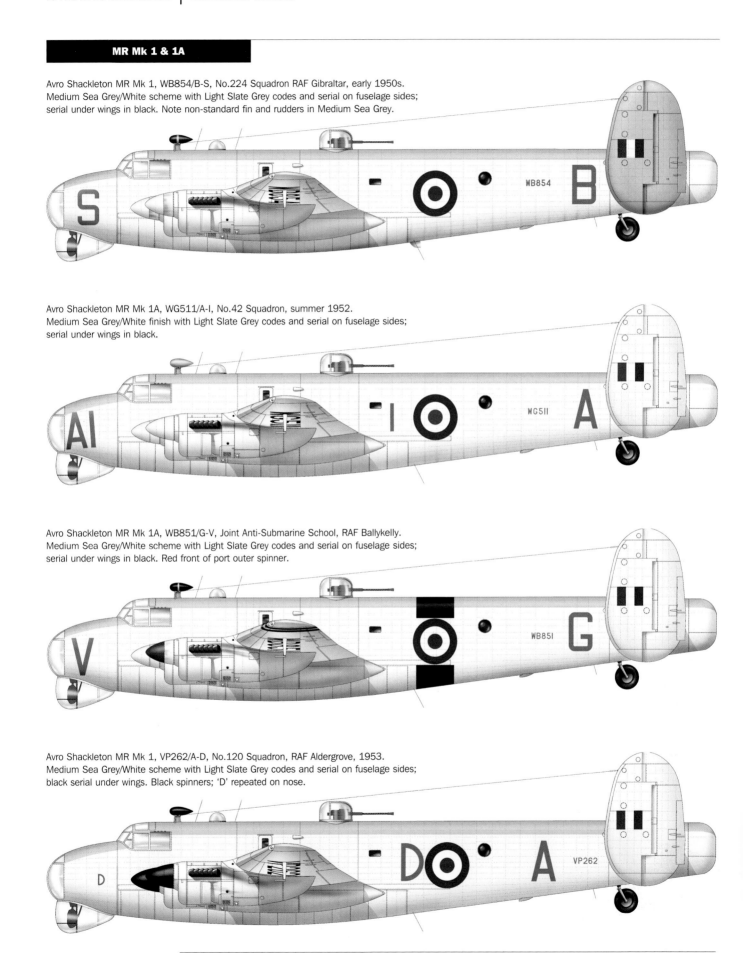

Avro Shackleton MR Mk 1, VP281/F-B, No.236 Operational Training Unit. Medium Sea Grey/White scheme with Light Slate Grey codes and serial on fuselage sides; serial under wings in black. Note 'B' repeated on nose.

Avro Shackleton MR Mk 1A, WB823/N, No.220 Squadron, RAF Bovingdon, 1956. Dark Sea Grey overall with red and white codes; red serial on rear fuselage, repeated under the wings outlined in white.

Avro Shackleton MR Mk 1A, WB854/H, No.120 Squadron, RAF Aldergrove, 1954. Dark Sea Grey overall with codes in red, outlined in white; serial in red, repeated under the wings, outlined in white. Note squadron crest on nose within a white disc.

Avro Shackleton MR Mk 1A, WB854/C, No.205 Squadron, RAF Changi, 1959. Dark Sea Grey overall with white fuselage top decking and front of spinners; red codes outlined in white. Red serial on rear fuselage, repeated under the wings outlined in white.

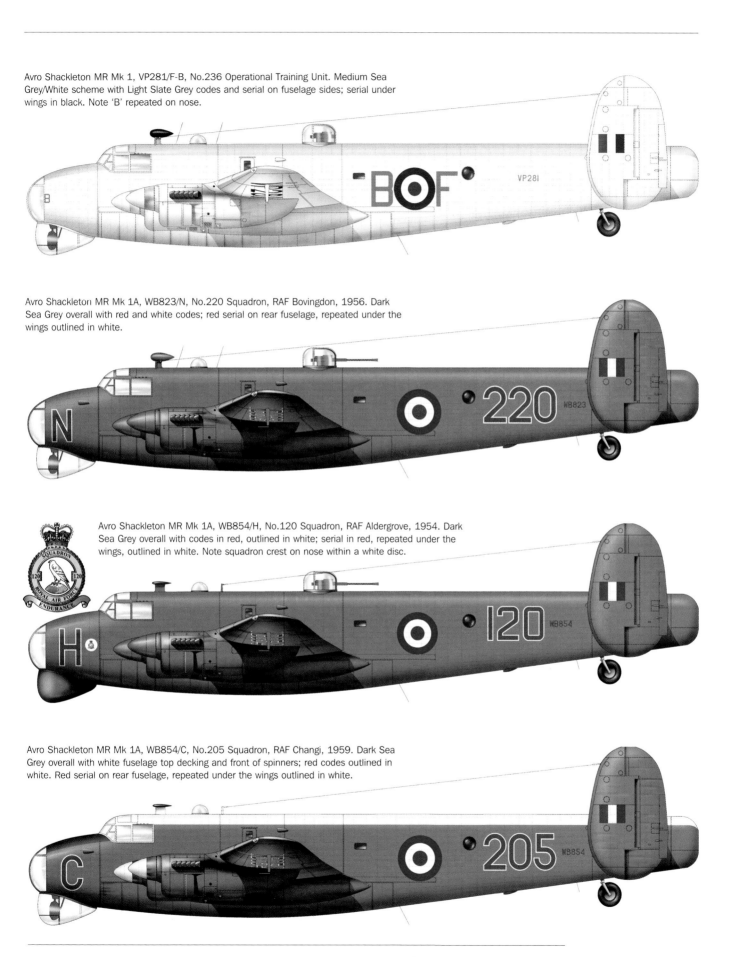

MR Mk 2

Avro Shackleton MR Mk 2 Phase II, WR961, 'S' of No. 38 Squadron based at
Malta, as marked during a visit to Ballykelly. General standard colour scheme of
Dark Sea Grey overall with white fuselage top decking and black anti-dazzle
panel. Red spinner fronts; rather unusual is the fuselage serial which is in red,
outlined in white similar to that under the wings. Codes in red, outlined white.
Union Jack on fin above fin flash and 'Royal Air Force' in white below; unit badge
is carried on both sides of the nose.

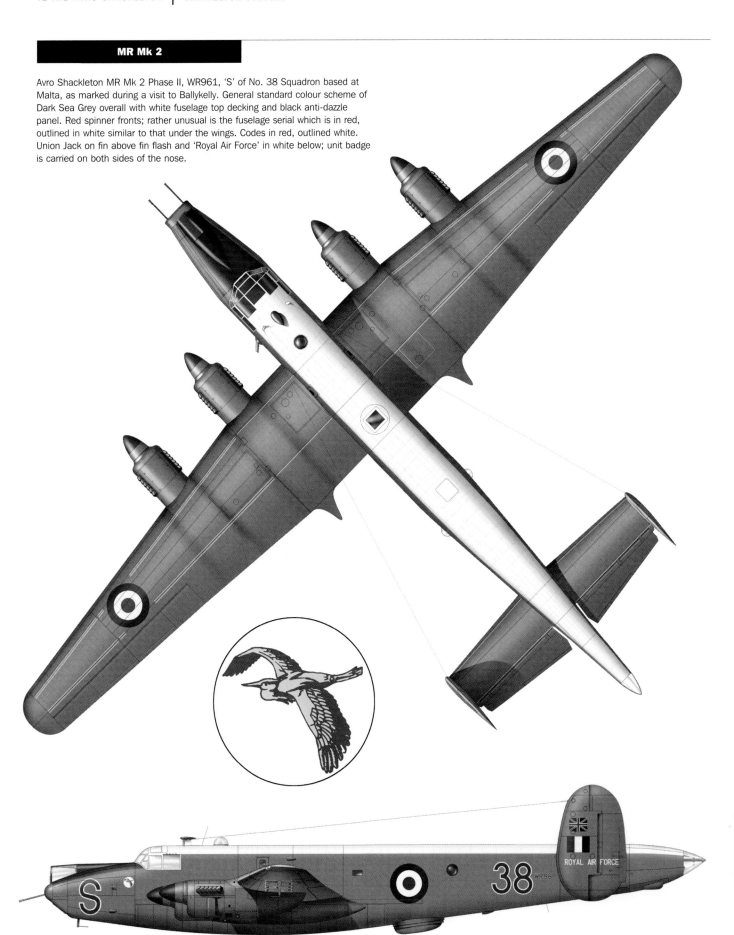

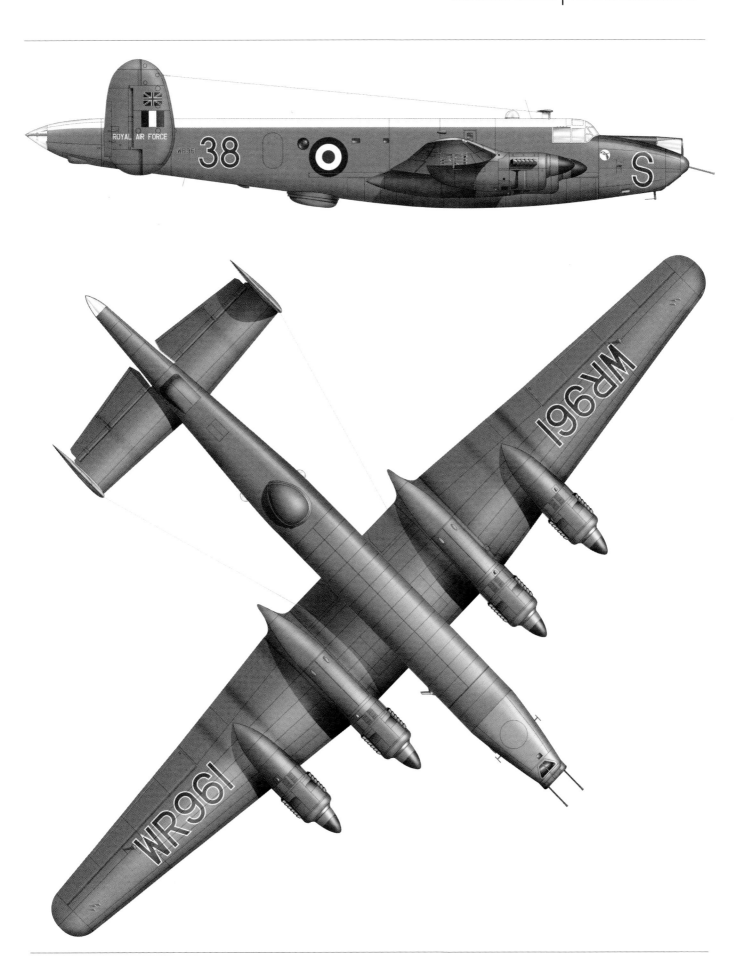

MR Mk 1

Avro Shackleton MR Mk 1, VP291, No.269 Squadron, 1958. Dark Sea Grey overall with white fuselage top decking; code in red, outlined in white. Red serial on fuselage sides, repeated under wings outlined in white; note no individual code letter.

MR Mk 2 & AEW Mk 2

Avro Shackleton MR Mk 2, WL789/F-D, Anti-Submarine Warfare Development Unit, late 1953. Medium Sea Grey/White scheme with Light Slate Grey codes and serial on fuselage; serial repeated in black under wings. Stripes on MAD boom and spinners in black.

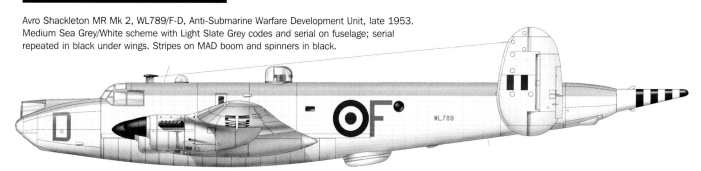

Avro Shackleton MR Mk 2, WG557/L, No.228 Squadron, RAF St. Eval, 1955. Medium Sea Grey/White scheme with red codes; serial on fuselage sides in Light Slate Grey, repeated under the wings in black.

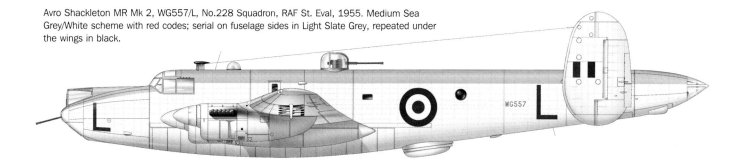

Avro Shackleton MR Mk 2, WL741/O, No.224 Squadron, during Exercise Medflex Champion, Oban, September 1955. Medium Sea Grey/White scheme with Light Slate Grey codes and serial on fuselage sides; serial repeated under wings in black. Unit badge on fin, above fin flash.

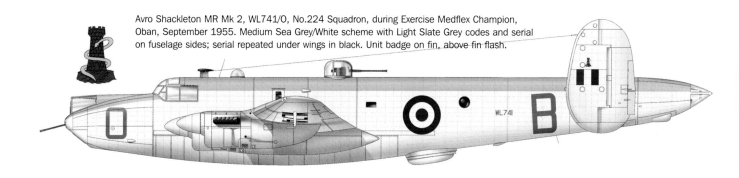

Avro Shackleton MR Mk 2, WL785/E, No.37 Squadron, RAF Luqa (Malta), 1956. Dark Sea Grey overall with yellow/black 'Suez' stripes around rear fuselage and wings; red codes outlined in white. Fuselage serial in red, repeated under the wings outlined in white.

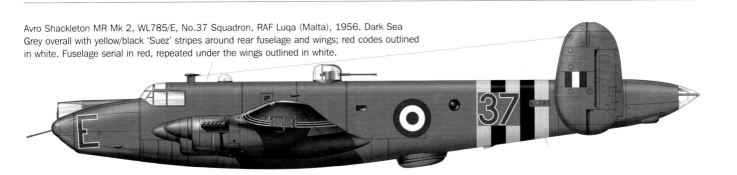

Avro Shackleton MR Mk 2, WL800/Z, No.37 Squadron, Malta, July 1957. Medium Sea Grey/White scheme with Light Slate Grey codes and serial on fuselage sides; serial repeated in black under wings. Spinners believed to be in Medium Sea Grey.

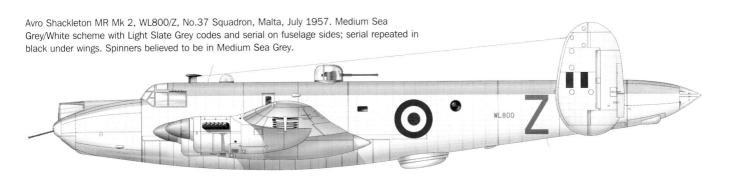

Avro Shackleton MR Mk 2, WR953/4, No.228 Squadron, during Goodwill Tour of South America, October-November 1955. Dark Sea Grey overall with red codes, outlined in red. Red serial on fuselage sides, repeated under the wings outlined in white. Union Jack on fin in place of usual fin flash; note white outlined roundels.

Avro Shackleton MR Mk 2 (Phase 2), WR964/Q, No.204 Squadron, RAF Ballykelly, May 1963. Dark Sea Grey overall with white fuselage top decking; codes in red, outlined in white. Serial on fuselage sides in red, repeated under the wing outlined in white; unit badge in a white shield on fin. Red front section of spinners.

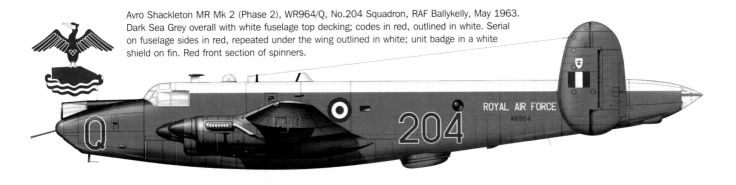

Avro Shackleton MR Mk 2, WR958/D, No.42 Squadron, RAF St Mawgan, January 1965. Dark Sea Grey overall with white fuselage top decking; codes in red, outlined in white. Serial on fuselage sides in red, repeated under the wings outlined in white; Union Jack on front fuselage and unit badge on fin.

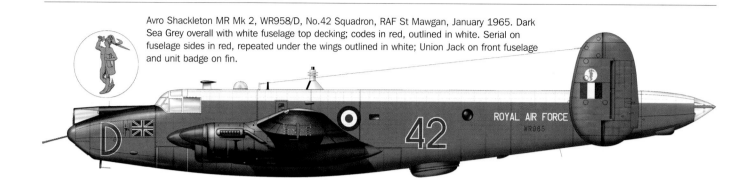

Avro Shackleton MR Mk 2, WL744/X, No.210 Squadron, RAF Ballykelly, 1963. Dark Sea Grey overall with white fuselage top decking; codes in red, outlined white. Serial in red on rear fuselage, repeated under the wings outlined in white; green spinners. Unit badge on forward fuselage.

Avro Shackleton MR Mk 2, WR967/Z, No.38 Squadron, RAF Hal Far (Malta), 1961. Dark Sea Grey overall with white fuselage top decking; red codes, outlined in white. Red spinners and serial on fuselage sides, the latter repeated under the wings, outlined in white; note Maltese Cross in Orange Day-glo on fin and unit badge on nose.

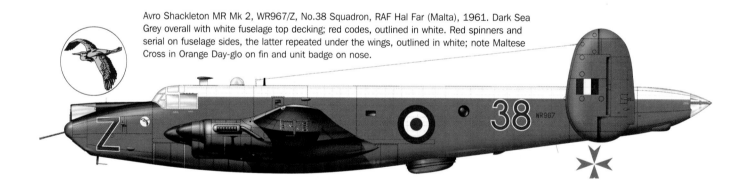

Avro Shackleton AEW Mk 2, WL757, No.8 Squadron, RAF Kinloss. Dark Sea Grey overall with black anti-dazzle panel; '8' on fuselage, '57' on fin and underwing serials in red, outlined in white. Red fuselage serial; roundel flashes in yellow, blue and red. Note rank pennant and unit badge on forward fuselage.

MR Mk 3

Avro Shackleton MR Mk 3, WR987/R, No.201 Squadron, seen on Malta in October 1958. Dark Sea Grey overall with red codes outlined in white; red serial on fuselage sides, repeated under wings outlined in white. Note unpainted exhaust stack panels.

Avro Shackleton MR Mk 3, XF710/K, No.210 Squadron, RAF St Mawgan, February 1962. Dark Sea Grey overall with white fuselage top decking; red wingtips and fixed part of horizontal tail surfaces. Red codes, outlined in white; red serial, repeated under the wings outlined in white. Unit badge in blue on a white shield on fin and Union Jack on nose.

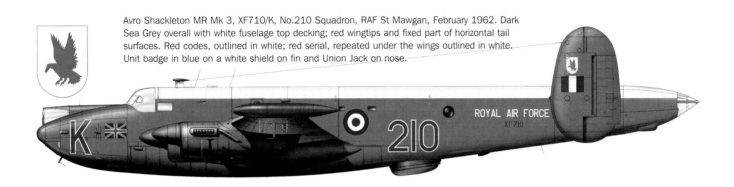

Avro Shackleton MR Mk 3 (Phase 2), WR971/E, No.210 Squadron, RAF Ballykelly. Dark Sea Grey overall with white fuselage top decking and black anti-dazzle panel; codes in red, outlined white. Red serial on rear fuselage, repeated under wing outlined in white; squadron crest on forward fuselage.

Avro Shackleton MR Mk 3 (Phase 3 without Viper engines), WR981/R, No.260 Squadron, as seen at RAE Llanbedr, June 1965. Dark Sea Grey overall with white fuselage top decking and black anti-dazzle panel; codes in red, outlined white. Serial in red on fuselage sides repeated under wings, outlined in white; unit badge on fin and Union Jack on forward fuselage.

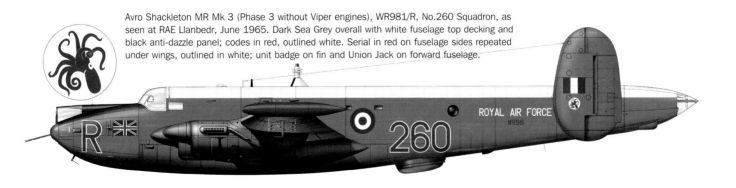

Avro Shackleton MR Mk 3 (Phase 3), WR988/E, No.203 Squadron, RAF Luqa (Malta), November 1971. Dark Sea Grey overall with white fuselage top decking and black anti-dazzle panel; red codes, outlined in white. Red serial, repeated under the wings outlined in white; unit crest on forward fuselage.

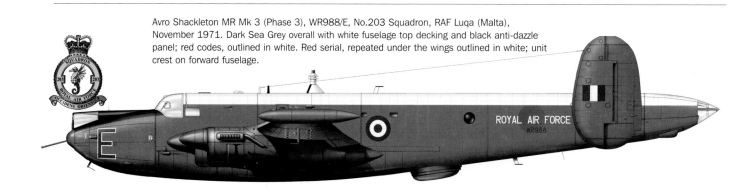

Avro Shackleton MR Mk 3, WR972, Royal Aircraft Establishment Farnborough, 1959. Aluminium uppersurfaces with yellow and black undersides; black anti-dazzle panel. Black serial, repeated under the wings.

T Mk 4

Avro Shackleton T Mk 4, VP293, Royal Aircraft Establishment Farnborough, 1964. Dark Sea Grey overall with white fuselage top decking; orange nose, fins/rudders and fuselage tail end, wing leading edges. Serial on fuselage in red, repeated under the wings in red outlined in white; inboard spinners in orange and white, outboard spinners in black/white halved front and orange rear. Black/white forward blades of outboard and rear blades of inboard propellers; roundels above wings outlined in white.

Avro Shackleton T Mk 4, VP259/Q, Maritime Operational Conversion Unit, 1957. Dark Sea Grey overall with black/white stripes around fuselage and wings; codes in red, outlined in white. Serial in black on rear fuselage, repeated in red, outlined in white, under the wings.

Avro Shackleton T Mk 4, WB844/L, Maritime Operational Training Unit, RAF St. Mawgan, mid-1960s. Dark Sea Grey overall with white fuselage top decking; red/white codes. Black serial under tailplane, repeated in red, outlined in white, under the wings; black/white badge on nose.

WB844

SAAF

Avro Shackleton MR Mk 3, 1717/O, No.35 Squadron, South African Air Force. Delivery finish of Dark Sea Grey upper surfaces and PR Blue undersides; roundels in six positions. Fin flashes on outer faces of fins only; serial in black, repeated under the wings. Only 1716, 1717 and 1718 were delivered with roundels. Note unit badge on nose.

Avro Shackleton MR Mk 3, 1717/O, No.35 Squadron, South African Air Force. Colour scheme as previous 1717, but with white fuselage top decking, black anti-dazzle panel and a rank pennant under the windscreen; yellow lifeboat with small national marking and black '3'. This type of lifeboat had been experimentally fitted to WR972 at A&AEE Boscombe Down.

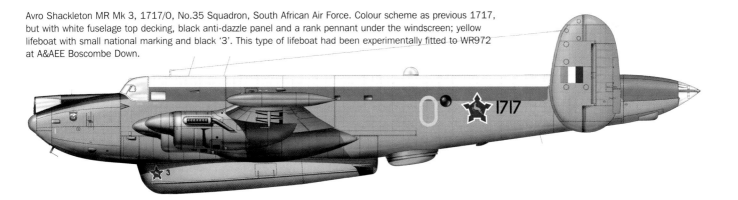

Avro Shackleton MR Mk 3, 1720/M, No.35 Squadron, South African Air Force. Dark Sea Grey on uppersurfaces and PR Blue undersides; white fuselage top decking and black anti-dazzle panel. Black serial on fuselage repeated under the wings; code in yellow. Red spinner fronts; unit badge on forward fuselage.

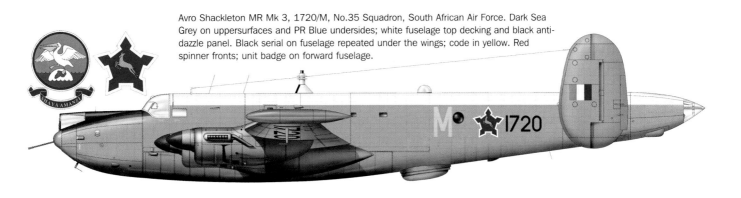

AEW Mk 2, WL793 was called 'Ermintrude' while in service with No.8 Squadron. This aircraft was eventually scrapped at Lossiemouth in July 1982.

No.8 Squadron AEW Mk 2, WL795, was called 'Rosalie' and today can be seen on the gate at RAF St. Mawgan, but is seen here during a visit to Malta.

Close-up of WL795, 'Rosalie' of No.8 Squadron on Malta during a visit there.
All R J Caruana

MR Mk 3, 1722 shown on the ramp at Ysterplaat after a flying display there on the 8th November 2003.
Glen Cameron-Williger

This official diagram shows
the wing markings on the
MR Mk 3.
Crown Copyright

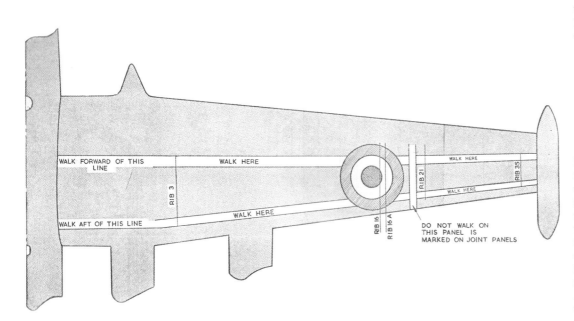

MR Mk 2, WR959 looking
rather the worst for wear
whilst with No.205
Squadron. The condition of
the airframe coupled with
the fact that No.205
Squadron was the last to
operate this machine
before it was reduced to
components at Changi in
1968, probably indicates
that this image was taken
not long before this aircraft
went out of service.
via Martyn Chorlton

Appendix I: **TECHNICAL DATA**

MARITIME RECONNAISSANCE

Designation: MR Mk 1

First Flight: 9th March 1949 (VW126)
Span: 120ft 0in
Length: 77ft 6in
Height: 17ft 6in
Tailplane Span: 33ft 0in
Engine: Four twelve-cylinder, 60' Vee, liquid-cooled, two-speed, single-stage supercharged Rolls-Royce Griffon Mk 57 (inboard) and Mk 57A (outboard) producing 1,960hp (2,435hp with water methanol) at maximum power.
Fuel Capacity: 3,292 Imp. Gal (+400 Imp. Gal. overload)
Weight: Basic 54,500lb, Landing 72,000lb, Max 86,000lb
Max speed @ 12,000ft: 294mph
Max cruise @ 10,000ft: 245mph
Landing speed: 114mph
Stalling speed: 88mph
Initial rate of climb: 1,005ft/min
Ceiling: 20,700ft
Range (nm) **: 2,160
Radius of 4hr patrol (nm): 780
Endurance (20% reserve): 14.8hrs
Max still air range (nm): 3,090
Take-off distance to 50 ft using Water Methanol: 3,750ft
Armament: Two 20mm Hispano cannon in mid-upper turret. Maximum bomb load of 14,000lb comprising various numbers of depth charges, mines, sonobuoys and 500lb and 1,000lb free-fall bombs
Propeller: De Havilland 13 ft diameter counter-rotating, constant-speed, fully feathering units.
  Front: left-hand tractor DF 1 60/3 34/1 or DB 171/334/1 with basic setting and fine pitch 23', feathered 90'.
  Rear: right-hand tractor DF 1 60/336/1 or DB 171/336/1 with basic setting and fine pitch 24', feathered 91'.
Production: 29

Designation: MR Mk 1A

First Flight: N/K
Span: 120ft 0in
Length: 77ft 6in
Height: 17ft 6in
Tailplane Span: 33ft 0in
Engine: Four twelve-cylinder, 60' Vee, liquid-cooled, two-speed, single-stage supercharged Rolls-Royce Griffon Mk 57A producing 1,960hp (2,435hp with water methanol) at maximum power.
Fuel Capacity: 3,292 Imp. Gal (+400 Imp. Gal. overload)
Weight: Basic 54,500lb, Landing 72,000lb, Max 86,000lb
Max speed @ 12,000ft: 294mph
Max cruise @ 10,000ft: 245mph
Landing speed: 114mph
Stalling speed: 88mph
Initial rate of climb: 1,005ft/min
Ceiling: 20,700ft
Range (nm) **: 2,160
Radius of 4hr patrol (nm): 780
Endurance (20% reserve): 14.8hrs
Max still air range (nm): 3,090
Take-off distance to 50 ft using Water Methanol: 3,750ft
Armament: Two 20mm Hispano cannon in mid-upper turret. Maximum bomb load of 14,000lb comprising various numbers of depth charges, mines, sonobuoys and 500lb and 1,000lb free-fall bombs
Propeller: De Havilland 13 ft diameter counter-rotating, constant-speed, fully feathering units.
  Front: left-hand tractor DF 1 60/3 34/1 or DB 171/334/1 with basic setting and fine pitch 23', feathered 90'.
  Rear: right-hand tractor DF 1 60/336/1 or DB 171/336/1 with basic setting and fine pitch 24', feathered 91'.
Production: 47

Designation: MR Mk 2 (inc Phase 1 & 2)

First Flight: 19th July 1951 (VW126)
Span: 120ft 0in
Length: 87ft 4in
Height: 16ft 9in
Tailplane Span: 33ft 0in
Engine: Four twelve-cylinder, 60' Vee, liquid-cooled, two-speed, single-stage supercharged Rolls-Royce Griffon Mk 57A producing 1,960hp (2,435hp with water methanol) at maximum power.
Fuel Capacity: 3,292 Imp. Gal (+400 Imp. Gal. overload)
Weight: Landing 72,000lb, Max 89,000lb
Max speed @ 12,000ft: 299mph
Max cruise @ 10,000ft: 249mph
Landing speed: 114mph
Stalling speed: 88mph
Initial rate of climb (ft/min): 920
Ceiling: 20,200ft
Range (nm) **: 1,980
Radius of 4hr patrol (nm): 700
Endurance (20% reserve): 14.6hrs
Max still air range (nm): 2,900
Take-off distance to 50 ft using Water Methanol: 3,750ft
Armament: Two 20mm Hispano cannon in mid-upper turret. Maximum bomb load of 14,000lb comprising various numbers of depth charges, mines, sonobuoys and 500lb and 1,000lb free-fall bombs.
Propeller: De Havilland 13 ft diameter counter-rotating, constant-speed, fully feathering units. Front: left-hand tractor DF 1 60/3 34/1 or DB 171/334/1 with basic setting and fine pitch 23', feathered 90'. Rear: right-hand tractor DF 1 60/336/1 or DB 171/336/1 with basic setting and fine pitch 24', feathered 91'.
Production: 70 (inc Phase 3 machines)

Designation: MR Mk 2 (Phase 3)

First Flight: N/K
Span: 120ft 0in
Length: 87ft 4in
Height: 16ft 9in
Tailplane Span: 33ft 0in
Engine: Four twelve-cylinder, 60' Vee, liquid-cooled, two-speed, single-stage supercharged Rolls-Royce Griffon Mk 58 producing 1,960hp (2,435hp with water methanol) at maximum power.
Fuel Capacity: 3,292 Imp. Gal (+400 Imp. Gal. overload)
Weight: Basic 59,000lb, Landing 72,000lb, Max 95,500lb
Max speed @ 12,000ft: 286mph
Max cruise @ 10,000ft: 249mph
Landing speed: 116mph
Stalling speed: 92mph
Initial rate of climb (ft/min): 900

Ceiling: 18,800ft
Range (nm) **: 1,720
Radius of 4hr patrol (nm): 670
Endurance (20% reserve): 13hrs
Max still air range (nm): 2,780
Take-off distance to 50 ft using Water Methanol: 5,460ft
Armament: Two 20mm Hispano cannon in mid-upper turret.
Maximum bomb load of 14,000lb comprising various numbers
of depth charges, mines, sonobuoys and 500lb and 1,000lb
free-fall bombs.
Propeller: De Havilland 13 ft diameter counter-rotating,
constant-speed, fully feathering units.
Front: left-hand tractor DF 1 60/3 34/1 or DB 171/334/1
with basic setting and fine pitch 23', feathered 90'.
Rear: right-hand tractor DF 1 60/336/1 or DB 171/336/1
with basic setting and fine pitch 24', feathered 91'.
Production: 70 (inc Phase 1 & 2 machines)

Designation: MR Mk 3 (inc Phase 1 & 2)

First Flight: 2nd September 1955 (WR970)
Span: 119ft 10in
Length: 87ft 4in
Height: 23ft 4in
Tailplane Span: 33ft 0in
Engine: Four twelve-cylinder, 60' Vee, liquid-cooled, two-speed,
single-stage supercharged Rolls-Royce Griffon Mk 57A
producing 1,960hp (2,435hp with water methanol) at
maximum power.
Fuel Capacity: 4,248 Imp. Gal (+400 Imp. Gal. overload)
Weight: Basic 57,800lb, Landing 86,000lb, Max 100,000lb
Max speed @ 12,000ft: 262mph
Landing speed: 116mph
Stalling speed: 92mph
Initial rate of climb (ft/min): 900
Endurance (20% reserve): 13hrs
Max still air range (nm): 2,700
Armament: Two 20mm Hispano cannon in mid-upper turret.
Maximum bomb load of 10,000lb comprising various numbers
of torpedoes, depth charges, mines, sonobuoys, marine
markers and 500lb and 1,000lb free-fall bombs.
Propeller: De Havilland 13 ft diameter counter-rotating,
constant-speed, fully feathering units.
Front: left-hand tractor DF 1 60/3 34/1 or DB 171/334/1
with basic setting and fine pitch 23', feathered 90'.
Rear: right-hand tractor DF 1 60/336/1 or DB 171/336/1
with basic setting and fine pitch 24', feathered 91'.
Production: 34 (inc Phase 3 machines)

Designation: MR Mk 3 Phase 3

First Flight: N/K
Span: 119ft 10in
Length: 87ft 4in
Height: 23ft 4in
Tailplane Span: 33ft 0in
Engine: Four twelve-cylinder, 60' Vee, liquid-cooled, two-speed,
single-stage supercharged Rolls-Royce Griffon Mk 58
producing 1,960hp (2,435hp with water methanol) at
maximum power. Also two Bristol-Siddeley Viper 203 jet
engines producing 2,500lb static thrust under the outboard
engine nacelles.
Fuel Capacity: 4,316 Imp. Gal (+400 Imp. Gal. overload)
Weight: Basic 64,300lb, Landing 88,000lb, Max 108,000lb
Max speed @ 12,000ft: 297mph
Max cruise @ 10,000ft: 253mph
Landing speed: 118mph
Stalling speed: 96mph
Initial rate of climb (ft/min): 850
Ceiling: 18,600ft
Range (nm) **: 2,300

Radius of 4hr patrol (nm): 970
Endurance (20% reserve): 16hrs
Max still air range (nm): 1,660
Take-off distance to 50ft using Water Methanol: 5,550ft
Armament: Two 20mm Hispano cannon in mid-upper turret.
Maximum bomb load of 10,000lb comprising various numbers
of torpedoes, depth charges, mines, sonobuoys, marine
markers and 500lb and 1,000lb free-fall bombs.
Propeller: De Havilland 13 ft diameter counter-rotating,
constant-speed, fully feathering units.
Front: left-hand tractor DF 1 60/3 34/1 or DB 171/334/1
with basic setting and fine pitch 23', feathered 90'.
Rear: right-hand tractor DF 1 60/336/1 or DB 171/336/1
with basic setting and fine pitch 24', feathered 91'.
Production: 34 (inc Phase 1 & 2 machines)

AIRBORNE EARLY WARNING

Designation: AEW Mk 2 (Phase 3)

First Flight: 30th September 1971 (WL745)
Span: 119ft 10in
Length: 87ft 4in
Height: 23ft 4in
Tailplane Span: 33ft 0in
Engine: Four twelve-cylinder, 60' Vee, liquid-cooled, two-speed,
single-stage supercharged Rolls-Royce Griffon Mk 58,
1,960hp (2,435hp with water methanol) at maximum power.
Armament: Maximum bomb load of 10,000lb comprising various
numbers of torpedoes, depth charges, mines, sonobuoys,
marine markers and Lindholme gear.
Propeller: De Havilland 13 ft diameter counter-rotating,
constant-speed, fully feathering units.
Front: left-hand tractor DF 1 60/3 34/1 or DB 171/334/1
with basic setting and fine pitch 23', feathered 90'.
Rear: right-hand tractor DF 1 60/336/1 or DB 171/336/1
with basic setting and fine pitch 24', feathered 91'.
Production: 12

TRAINERS

Designation: T Mk 2

Span: 119ft 10in
Length: 87ft 4in
Height: 23ft 4in
Tailplane Span: 33ft 0in
Engine: Four twelve-cylinder, 60' Vee, liquid-cooled, two-speed,
single-stage supercharged Rolls-Royce Griffon Mk 58,
1,960hp (2,435hp with water methanol) at maximum power.
Also two Bristol-Siddeley Viper 203 jet engines producing
2,500lb static thrust under the outboard engine nacelles.
Weight: Max (AUW) lb (kg)
Max speed @ 12,000ft: 262mph
Landing speed: 116mph
Stalling speed: 92mph
Initial rate of climb (ft/min): 900
Endurance (20% reserve): 13hrs
Max still air range (nm): 2,700
Armament: Maximum bomb load of 14,000lb comprising various
numbers of torpedoes, depth charges, mines, sonobuoys,
marine markers and 500lb and 1,000lb free-fall bombs.
Propeller: De Havilland 13 ft diameter counter-rotating,
constant-speed, fully feathering units.
Front: left-hand tractor DF 1 60/3 34/1 or DB 171/334/1 with
basic setting and fine pitch 23', feathered 90'.
Rear: right-hand tractor DF 1 60/336/1 or DB 171/336/1 with
basic setting and fine pitch 24', feathered 91'.
Production: 10 (converted from MR Mk 2s)

Designation: T Mk 4

Span: 119ft 10in
Length: 87ft 4in
Height: 23ft 4in
Tailplane Span: 33ft 0in
Engine: Four twelve-cylinder, 60' Vee, liquid-cooled, two-speed, single-stage supercharged Rolls-Royce Griffon Mk 57 (inboard) and 57A (outboard) producing 1,960hp (2,435hp with water methanol) at maximum power.
Fuel Capacity: 3,292 Imp. Gal (+400 Imp. Gal. overload)
Weight: Max (AUW) lb (kg)
Max speed @ 12,000ft: 294mph
Max cruise @ 10,000ft: 245mph
Landing speed: 114mph
Stalling speed: 88mph
Initial rate of climb: 1,005ft/min
Ceiling: 20,700ft
Range (nm) **: 2,160
Radius of 4hr patrol (nm): 780
Endurance (20% reserve): 14.8hrs
Max still air range (nm): 3,090
Take-off distance to 50 ft using Water Methanol: 3,750ft
Armament: Maximum bomb load of 14,000lb comprising various numbers of torpedoes, depth charges, mines, sonobuoys marine markers and 500lb and 1,000lb free-fall bombs
Propeller: De Havilland 13 ft diameter counter-rotating, constant-speed, fully feathering units.
 Front: left-hand tractor DF 1 60/3 34/1 or DB 171/334/1 with basic setting and fine pitch 23', feathered 90'.
 Rear: right-hand tractor DF 1 60/336/1 or DB 171/336/1 with basic setting and fine pitch 24', feathered 91'.
Production: 17 (converted from MR Mk 1s)

* Mk 3 Phase III (Viper) had improved take-off performance (4,750 ft to clear 50ft screen), but reduced range and endurance.
** @5,000 ft at 190mph (full load - 20% reserve)

Note: Performance figures need to be considered with caution because it is extremely difficult to ensure that the same parameters have been applied by all the sources. For instance 'normal' range will have an allowance built in for diversion and/or contingencies and this allowance may be either up to 20% of the total fuel, 20% of the 'off-task' fuel (in the case of maritime operations), or as in the case of some Avro calculations for the Shackleton, 10% of the overall fuel. Unless otherwise stated, range and endurance speeds/times include a 20% reserve of total fuel.

We have refrained from using metric conversions on the above tables, as the Shackelton was built to Imperial specifications and dimensions and any conversion to metric would require working to many decimal places to be anywhere near accurate, so we have omitted any mention of metric at all.

CONVERSION CHART

Imperial
1 nautical mile (nm) = 1. 136 statute miles
1 miles per hour (mph) = 0.88 knots
1 Imp Gallon (Imp. Gal.) = 7.2 lb
Imperial to Metric
1 foot = 0.3048 metres
1 nautical mile = 1.828 kilometres
1 square foot = 0.0929 square metres
1 ft per minute = 0.005 m/sec
1mph = 1.61 km/h
1lb = 0.454 kg
1 Gallon = 4.546 litres

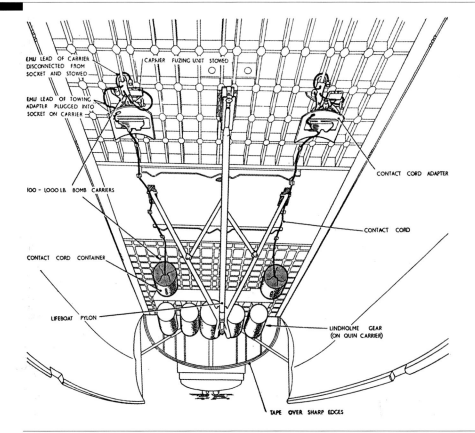

This diagram from the official publication of the Airborne Lifeboat gives a good idea of the equipment in the Shackleton bomb bay for the ASR role.
Crown Copyright

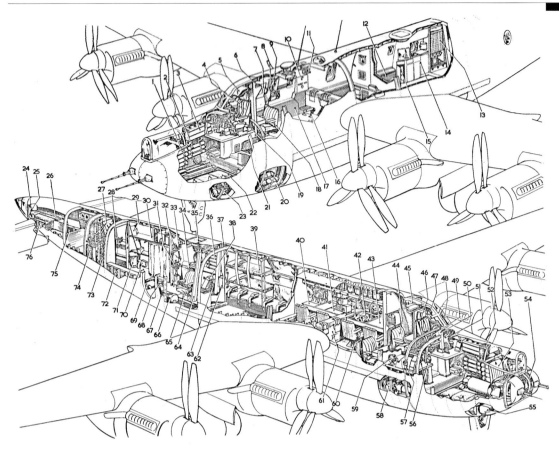

A general view of the equipment in the fuselage of the MR Mk 3. *Crown Copyright*

Key

1. Stowage (Special Stores)
2. Stowage (Marine Marker)
3. Ammunition system (Port and Starboard)
4. Second pilot's seat
5. Canopy control panel
6. Signal pistol stowage
7. Signal cartridge stowage
8. Drift recorder
9. Air thermometer
10. First aid kit and axe
11. Periscopic sextant
12. Parachute dinghy stowage
13. Flare discharger station
14. Galley
15. Waste bin
16. Step
17. Main power panels
18. Flight Engineer's seat
19. Parachute exit release
20. Nitrogen bottles
21. Parachute escape hatch
22. Air bomber's cushions
23. Relief tank
24. Wander light
25. Astro compass
26. Rear Observer's cushions
27. Flying controls
28. Flying control guard rail
29. Rudder locking system
30. Wardrobe
31. Drift sight
32. Binocular stowage

33. Astro compass stowage
34. Port observer's seat
35. Escape hatch
36. Reconnaissance flare stowage
37. Signal cartridge discharger
38. Flare chute
39. Rest bunks
40. Radar operator's station
41. Sonobuoy operator's station
42. Tactical navigator's station
43. Routine navigator's station
44. Signaller's seat
45. Pilot's escape hatch
46. First pilot's seat
47. Crash handle
48. Windscreen wipers
49. Instrument panel
50. Control column
51. Throttle box
52. Ammunition system
53. Gunner's seat
54. Gun installation
55. Windscreen de-icing tank
56. Rudder controls
57. Aldis lamp stowage
58. No.2 heater
59. Trim control
60. Sextant stowage
61. Crash station
62. Flare stowage
63. Stowage (Special Stores)
64. Scanner unit
65. Starboard observer's seat

66. Portable dinghy
67. Hand camera stowage
68. Ditching rope
69. No.3 heater
70. First aid box
71. Elsan closet installation
72. Parachute and dinghy stowage
73. Washbasin
74. Relief tank
75. Vertical camera
76. Oblique camera

The mid-section of the fuselage clearly showing the radome location on the MR Mk 3.

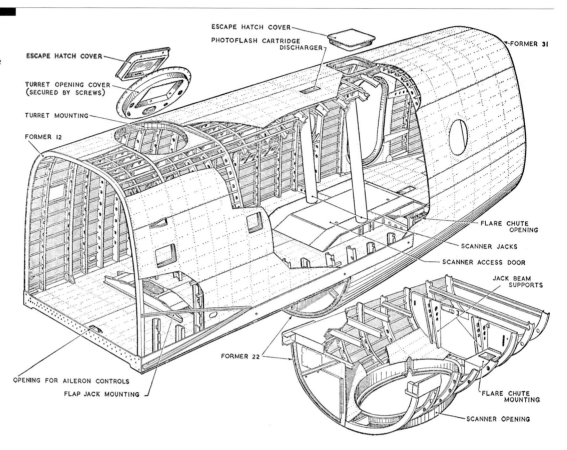

This diagram shows the inboard flaps of the MR Mk 3.
Both Crown Copyright

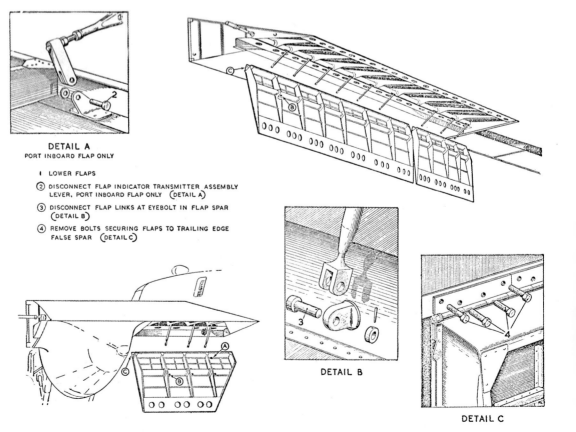

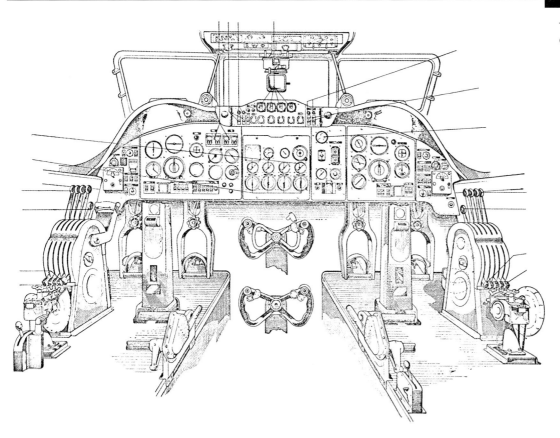

The main instrument panel of the MR Mk 3.

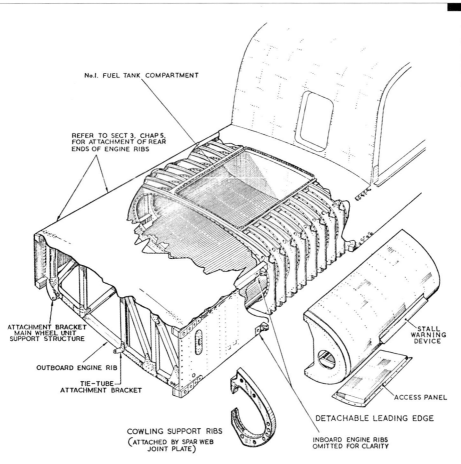

No.I. FUEL TANK COMPARTMENT

REFER TO SECT 3, CHAP 5,
FOR ATTACHMENT OF REAR
ENDS OF ENGINE RIBS

ATTACHMENT BRACKET
MAIN WHEEL UNIT
SUPPORT STRUCTURE

OUTBOARD ENGINE RIB

TIE-TUBE
ATTACHMENT BRACKET

COWLING SUPPORT RIBS
(ATTACHED BY SPAR WEB
JOINT PLATE)

INBOARD ENGINE RIBS
OMITTED FOR CLARITY

DETACHABLE LEADING EDGE

ACCESS PANEL

STALL
WARNING
DEVICE

The mid-wing section of the MR Mk 3. Note the stall warning strips on the leading edge.
Both Crown Copyright

The nose structure of the MR Mk 3.

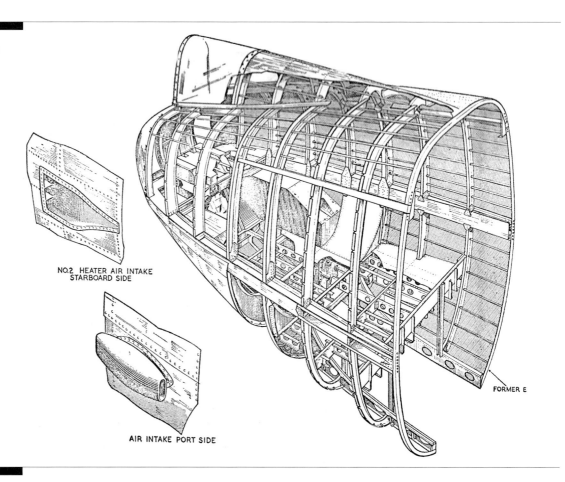

NO.2 HEATER AIR INTAKE
STARBOARD SIDE

AIR INTAKE PORT SIDE

FORMER E

The rear fuselage of the MR Mk 3. Note the tactical camera installation.
Both Crown Copyright

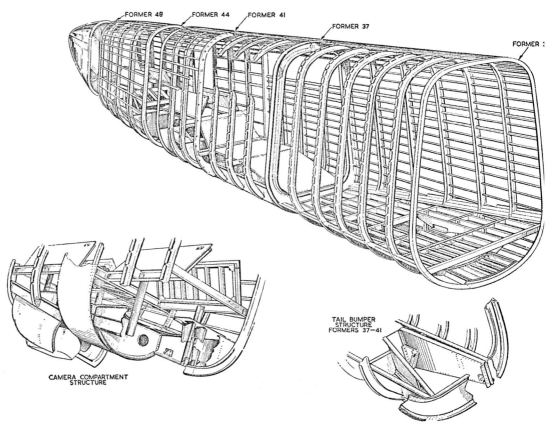

FORMER 48   FORMER 44   FORMER 41   FORMER 37   FORMER 3

CAMERA COMPARTMENT
STRUCTURE

TAIL BUMPER
STRUCTURE
FORMERS 37–41

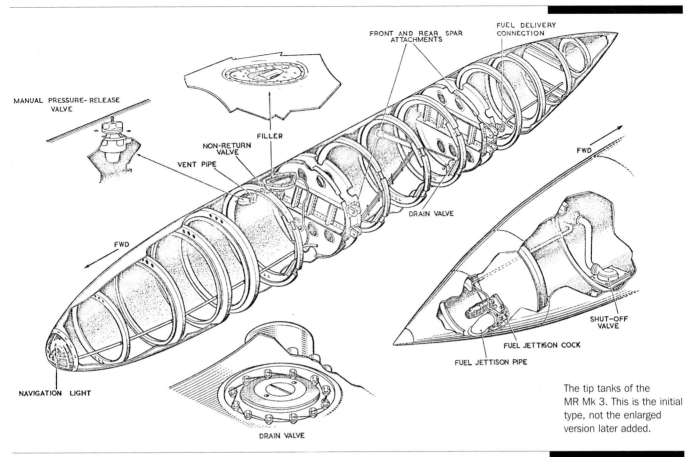

MANUAL PRESSURE-RELEASE VALVE

FILLER

NON-RETURN VALVE

VENT PIPE

FRONT AND REAR SPAR ATTACHMENTS

FUEL DELIVERY CONNECTION

FWD

FWD

DRAIN VALVE

NAVIGATION LIGHT

DRAIN VALVE

SHUT-OFF VALVE

FUEL JETTISON COCK

FUEL JETTISON PIPE

The tip tanks of the MR Mk 3. This is the initial type, not the enlarged version later added.

The main undercarriage unit of the MR Mk 3. *Both Crown Copyright*

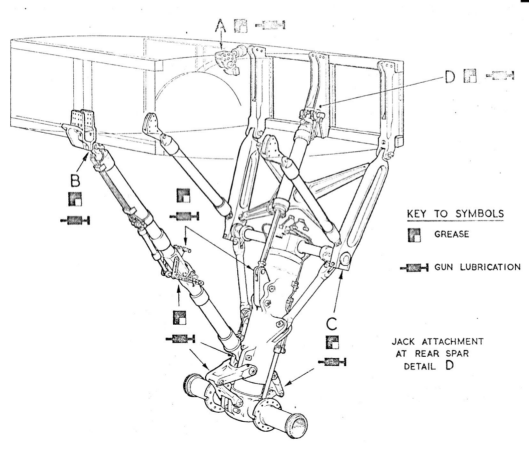

A

D

B

KEY TO SYMBOLS

GREASE

GUN LUBRICATION

C

JACK ATTACHMENT AT REAR SPAR DETAIL D

Useful view, front and
back, of the undercarriage
bay in the MR Mk 3.

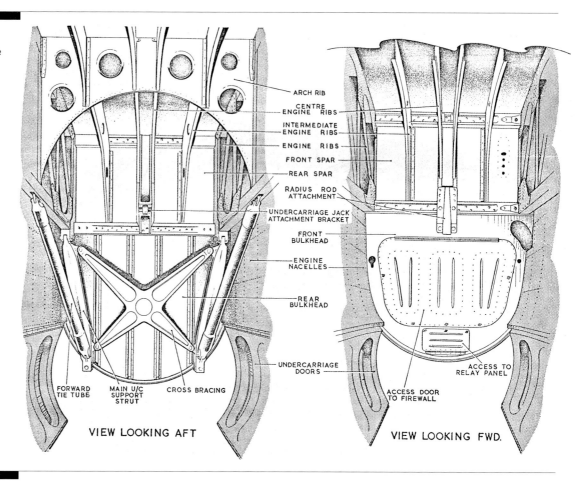

ARCH RIB

CENTRE
ENGINE RIBS

INTERMEDIATE
ENGINE RIBS

ENGINE RIBS

FRONT SPAR

REAR SPAR

RADIUS ROD
ATTACHMENT

UNDERCARRIAGE JACK
ATTACHMENT BRACKET

FRONT
BULKHEAD

ENGINE
NACELLES

REAR
BULKHEAD

UNDERCARRIAGE
DOORS

ACCESS TO
RELAY PANEL

ACCESS DOOR
TO FIREWALL

FORWARD
TIE TUBE

MAIN U/C
SUPPORT
STRUT

CROSS BRACING

VIEW LOOKING AFT

VIEW LOOKING FWD.

The Bristol Type B.17
mid-upper turret as used
in the Lincoln and
Shackleton
Both Crown Copyright

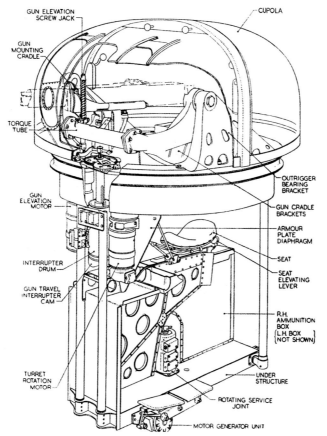

GUN ELEVATION
SCREW JACK

CUPOLA

GUN
MOUNTING
CRADLE

TORQUE
TUBE

GUN
ELEVATION
MOTOR

INTERRUPTER
DRUM

GUN TRAVEL
INTERRUPTER
CAM

TURRET
ROTATION
MOTOR

OUTRIGGER
BEARING
BRACKET

GUN CRADLE
BRACKETS

ARMOUR
PLATE
DIAPHRAGM

SEAT

SEAT
ELEVATING
LEVER

R.H.
AMMUNITION
BOX
[L.H. BOX
NOT SHOWN]

UNDER
STRUCTURE

ROTATING SERVICE
JOINT

MOTOR GENERATOR UNIT

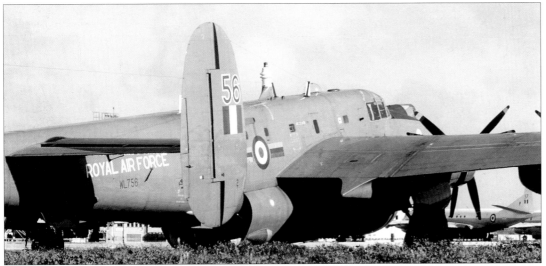

Top left: Close-up of the tail markings of AEW Mk 2, WL745 from No.8 Squadron.
R J Caruana

Top right: AEW Mk 2, WL745 from No.8 Squadron seen parked out on Malta during a visit there.
R J Caruana

Centre: This rear view of AEW Mk 2, WL756, '56' of No.8 Squadron, shows the aerial unit and leads as well as the large bomb bay doors (note the fairing on the rear of these). This photograph was taken on the 20th February 1977.
Godfrey Mangion

Above left: Useful view of the rear end of the MR Mk 3, note the tactical camera position. This is MR Mk 3, 'E' of No.203 Squadron photographed on the 28th November 1971. *Godfrey Mangion*

Above right: A useful view of the engines and mid-fuselage on the MR Mk 3. The air doors for the outboard Viper are down and the large antenna on the dorsal spine is installed. This is MR Mk 3, 'E' of No.203 Squadron photographed on the 28th November 1971.
Godfrey Mangion

A good view of the mid-section of AEW Mk 2, WL756, '56' of No.8 Squadron photographed on the 20th February 1977. *Godfrey Mangion*

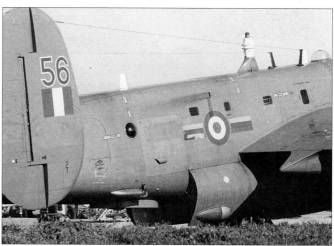

Centre left: Nice detail shot from the rear of AEW Mk 2, WR960 of No.8 Squadron. Note the massive aerial on the dorsal spine, the aerial leads and the blade antenna as well as the shape of the bomb bay doors.

Centre right: Close-up of the massive radome on WR960, a No.8 Squadron AEW Mk 2. Note the squadron crest on the support for the antenna.

Above left: An unidentified AEW Mk 2 of No.8 Squadron on the apron while visiting Malta. Note the mass of rivets and the reinforcing ribs on the bomb doors.

Above right: Nice detail view of the aft fuselage on AEW Mk 2, WL756 seen here in February 1977. Note the old ventral radome ring, aerials and leads and the rod antenna on the bomb bay doors.
All Godfrey Mangion

GUARDIAN OF THE SEA LANES

The Navy's Private Eye

Vast **stretches of ocean** . . . hour after hour of unceasing vigilance . . . climatic changes from tropical to antarctic . . . these are the conditions of a routine patrol to keep the sea lanes safe for sea-borne traffic. A vital task demanding rugged strength and endurance : two of the many outstanding qualities of the long range Avro Shackleton Mark 3. With its four Rolls Royce Griffon engines and its $1\frac{1}{2}$ tons of sensitive search and detection apparatus, this great aircraft is admirably suited to Maritime Reconnaissance. It can undertake very long periods of duty, and has even a large comfortable wardroom with an all-electric galley, where pre-cooked meals can be prepared for its crew of ten.

The Shackleton Mark 3 is now in production for R.A.F. Coastal Command and the South African Air Force. For this versatile aircraft is also an air weapon which can be used as an integral part of defence plants to protect and support sea-borne commerce. Its search and striking power make it the world's most effective submarine killer. It can also carry out vital air-sea rescue work to shipwrecked seamen. And a few squadrons are all that is needed for continuous day and night patrols over the countless miles of the Nation's sea lanes.

AVRO **Shackleton**

MARK 3

A. V. ROE & CO. LIMITED / MANCHESTER • ENGLAND   *Member of the Hawker Siddeley Group | Pioneer . . . and World Leader in Aviation*

Appendix II: SHACKLETON SQUADRONS

NO.8 SQUADRON

Started Shackleton Operations: Kinloss
    (1st January 1972)
UK Based: Kinloss (1/72-8/73), Lossiemouth (8/73-7/91)
Foreign Based: N/A
Reformed: N/A **Disbanded:** N/A
Re-equipped: 1st July 1991 (Boeing E-3A AWAC)
Variants Operated: AEW Mk 2

NO.37 SQUADRON

Started Shackleton Operations: Luqa, Malta
    (July 1953)
UK Based:
Foreign Based: Luqa, Malta (7/53-7/57) with
    detachments to Gibraltar, Trucial Oman, Durban,
    Khormaksar, Aden (7/57-9/67) with detachment to
    Majunga.
Reformed: N/A
Disbanded: 7th September 1967 (Khormaksar)
Re-equipped: N/A
Variants Operated: MR Mk 2

NO.38 SQUADRON

Started Shackleton Operations: Luqa, Malta
    (18th September 1953)
UK Based: N/A
Foreign Based: Luqa, Malta (9/53-9/65), Hal Far,
    Malta (9/65-3/67)
Reformed: N/A
Disbanded: 31st March 1967 (Hal Far)
Re-equipped: N/A
Variants Operated: MR Mk 2

NO.42 SQUADRON

Started Shackleton Operations: St Eval
    (28th June 1952)
UK Based: St. Eval (6/52-10/58), St. Mawgan
    (10/58-9/71) detachment to Aldergrove
Foreign Based: Detachments to Khormaksar & Masirah
    (6/52-10/58) and Gibraltar, Belize, Khormaksar,
    Masirah, Tengah, Majunda & Luqa (10/58)
Reformed: N/A **Disbanded:** N/A
Re-equipped: September 1971 (H.S. Nimrod MR Mk 1)
    Variants Operated: MR Mk 1A, MR Mk 2 & MR Mk 3
    Phase 3

NO.120 SQUADRON

Started Shackleton Operations: Kinloss (March 1951)
UK Based: Kinloss (3/51-4/52), Aldergrove (4/52-4/59),
    Kinloss (4/59-2/71)
Foreign Based: Detachment to Sola
Reformed: N/A **Disbanded:** N/A
Re-equipped: February 1971 (H.S. Nimrod MR Mk 1)
Variants Operated: MR Mk 1 & 1A, MR Mk 2 &
    MR Mk 3

NO.201 SQUADRON

Started Shackleton Operations: St. Mawgan (10/58)
UK Based: St. Mawgan (10/58- 7/65), Kinloss
    (7/65-12/70)
Foreign Based: N/A
Reformed: N/A **Disbanded:** N/A
Re-equipped: December 1970 (H.S. Nimrod MR Mk 1)
Variants Operated: MR Mk 1 & MR Mk 3

NO.203 SQUADRON

Started Shackleton Operations: Ballykelly
    (1st November 1958)
UK Based: Ballykelly (11/58-2/69)
Foreign Based: Luqa, Malta (2/69-10/71)
Reformed: N/A **Disbanded:** N/A
Re-equipped: October 1971 (H.S. Nimrod MR Mk 1)
Variants Operated: MR Mk 1A, MR Mk 2, MR Mk 3 &
    MR Mk 3 Phase 3

NO.204 SQUADRON

Started Shackleton Operations: Ballykelly
    (1st January 1954)
UK Based: Ballykelly (1/54-4/71), Honington (4/71-4/72)
Foreign Based: Detachment to Seletar & Majunga
Reformed: N/A
Disbanded: 28th June 1972 (Honington)
Re-equipped: N/A
Variants Operated: MR Mk 1A & MR Mk 2

NO.205 SQUADRON

Started Shackleton Operations: Changi (May 1958)
    UK Based: N/A
Foreign Based: Changi (5/58-10/71) with detachment
    to Seletar & Gan
Reformed: N/A
Disbanded: 31st October 1971 (Changi)
Re-equipped: N/A
Variants Operated: MR Mk 1A & MR Mk 2

NO.206 SQUADRON

Started Shackleton Operations: St. Eval
    (27th September 1952)
UK Based: St. Eval (9/52-1/58), St. Mawgan (1/58-7/65),
    Kinloss (7/65-10/70)
Foreign Based: N/A
Reformed: N/A **Disbanded:** N/A
Re-equipped: October 1970 (H.S. Nimrod MR Mk 1)
Variants Operated: MR Mk 1A, MR Mk 2 & MR Mk 3

NO.210 SQUADRON

Started Shackleton Operations: Ballykelly
    (1st December 1958)
UK Based: Ballykelly (12/58-10/70)
Foreign Based: Detachment to Majunda & Sharjah
    (11/70-11/71)
Reformed: 1st November 1970 (Sharjah)
Disbanded: 31st October 1970 (Ballykelly) &
    15th November 1971 (Sharjah)
Re-equipped: N/A
Variants Operated: MR Mk 2

NO.220 SQUADRON

Started Shackleton Operations: Kinloss
(24th September 1951)
UK Based: Kinloss (9/51-11/51), St. Eval
(11/51-3/57), St. Mawgan (3/57-10/58)
Foreign Based: N/A
Reformed: N/A
Disbanded: Renumbered as No.201 Squadron,
1st October 1958
Re-equipped: N/A
Variants Operated: MR Mk 1 & 1A, MR Mk 2 &
MR Mk 3

NO.224 SQUADRON

Started Shackleton Operations: Gibraltar (3/52)
UK Based: N/A
Foreign Based: Gibraltar (3/52-10/66) detachment
to Masirah.
Reformed: N/A
Disbanded: 31st October 1966 (Gibraltar)
Re-equipped: N/A
Variants Operated: MR Mk 1 & MR Mk 2

NO.228 SQUADRON

Started Shackleton Operations: St. Eval (1st July 1954)
UK Based: St. Eval (7/54-11/56), St. Mawgan
(11/56-1/58), St. Eval (1/58-3/59)
Foreign Based: N/A
Reformed: N/A
Disbanded: 6th March 1959 (St. Eval)
Re-equipped: N/A
Variants Operated: MR Mk 2

NO.240 SQUADRON

Started Shackleton Operations: Aldergrove
(1st May 1952)
UK Based: Aldergrove (5/52-5/52), St. Eval (5/52-6/52),
Ballykellly (6/52-11/58)
Foreign Based: N/A
Reformed: N/A
Disbanded: Renumbered as No.203 Squadron,
1st November 1958 (Ballykelly)
Re-equipped: N/A
Variants Operated: MR Mk 1, 1A & MR Mk 2

NO.269 SQUADRON

Started Shackleton Operations: Gibraltar
(1st January 1952)
UK Based: Ballykelly (3/52-12/58)
Foreign Based: Gibraltar (1/52-3/52), detachment to
Bodo & Christmas Island.
Reformed: N/A
Disbanded: Renumbered as No.210 Squadron,
1st December 1958 (Ballykelly)
Re-equipped: N/A
Variants Operated: MR Mk 1, 1A & MR Mk 2

Other Units which have used the Shackleton
Royal Air Force - Others

No. 236 OCU

Formed at Kinloss on the 31st July 1945.
Received first Shackleton on the 31st May 1951
Based at Kinloss from 1951 to 1st October 1956 when it
was combined with No.1 MRS to become the Maritime
Operational Training Unit (MOTU).
Operated MR Mk 1 & 1A

MOTU

Formed at Kinloss on the 1st October 1956 by
redesignating No.236 OCU. Based at Kinloss from October
1956 to July 1965. Moved to St. Mawgan in July 1965.
Last Shackleton training course finished on the 28th July
1970. Shackletons all departed by 21st December 1970.
Operated MR Mk 1 & 1A, MR Mk 2 & T Mk 2 & 4

ASWDU

Based at St. Mawgan from 10th May 1951.
Started Shackleton operations in March 1952. Unit moved
to Ballykelly at some time in 1955 and remained there
until disbanded on the 1st April 1970. Operated MR Mk 1
& 1A, MR Mk 2 & MR Mk 3

JASS (Joint Anti-submarine School)

Originally formed at Eglinton on the 30th January 1947.
Started Shackleton operations on the 18th March 1952 at
Ballykelly and remained there until March 1957 when it
disbanded and its remaining Shackletons were distributed
between the squadrons at St. Eval.
Operated MR Mk 1A and MR Mk 2

RAF Flying College, Manby

Operated MR Mk 1 & 1A and MR Mk 2

School of Maritime Reconnaissance

Operated T Mk 4

Servicing Units

No.5 MU, RAF Kemble, Gloucestershire
No.12 MU, RAF Kirkbride, Cumbria
No.23 MU, RAF Aldergrove, County Antrim,
Northern Ireland
No.27 MU, RAF Shawbury, Shropshire
No.32 MU, RAF St. Athan, South Glamorgan
No.38 MU, RAF Llandow, South Glamorgan
No.49 MU, RAF Colerne, Wiltshire
No.60 MU, RAF Leconfield, Humberside
No.71 MU, RAF Bicester, Oxfordshire
No.103 MU, RAF Akrotiri, Cyprus
No.137 MU, RAF Safi, Malta
No.390 MU, RAF Seletar, Singapore
ASF (Station Flight) Eastleigh, Kenya
CAPMF St. Mawgan, Cornwall
CCMC St. Mawgan, Cornwall

Government & Research

Telecommunications Flying Unit
Operated MR Mk 1A

RAE Farnborough (inc ETPS)

Operated MR Mk 2 WB833 & WG557, MR Mk 2 WR972
and WR974 and T Mk 4 VP293

A&AEE Boscombe Down

See Appendix IV

Foreign Service

South African Air Force (SAAF)
No.35 Squadron, D.F. Malan airfield, Cape Town
(1967-1984)
MR Mk 3 S/Nos. 1716 to 1723

Appendix III: **SHACKLETON PRODUCTION**

Specification Number:
Air Ministry 5/46
Contract Number:
6/ACFT.1077/CB6(a), dated 28th May 1947
Shackleton Prototypes:
Quantity: 3

VW126 First flight 9/03/49. Became aerodynamic test airframe for MR Mk 2, became 7626M at No.2 Radio School, broken up at RAF Yatesbury in 1965.

VW131 First flight 2/09/49. Used as test airframe for installation of Napier Nomad. Fuselage used for ditching trials at Bracebridge Heath, SOC 10/05/62

VW135 First flight 29/03/50. A&AEE, RAE & No.49 MU, Broken up for spares and SOC 13/04/54.

Specification Number:
Air Ministry 42/46
Contract Number:
6/ACFT.6062/CB6(a), dated 21st March 1946
Shackleton MR Mk 1.
Quantity: 30 but only 29 actually built
VP254 to 268.

VP254 First flight 28/03/50, IFF10 & SARAH trials, A&AEE and FEAF. Crashed South China Sea 9/12/58.

VP255 First flight 30/06/50. DFU, A&AEE, TFU, No.240 & 269 Sqns, Nos.23 & 5 MU, No.205 Sqn, No.23 MU, A&AEE, No.23 MU, sold for scrap 22/08/63.

VP256 First flight 18/09/50, RAF Manby, No.38 MU, Nos. 224 & 269 Sqns, Cat. 4 damage, SOC 14/04/55, scrapped at No.23 MU February 1963.

VP257 First flight 28/08/50. CSDE, No.38 MU, No.220 Sqn, No.49 MU, No.220 Sqn, No.23 MU, NEA, sold for scrap 28/02/63.

VP258 First flight 13/10/50, No.38 MU, No.120 Sqn, converted to T Mk 4 (5/55), A&AEE, MoS Air Fleet, MOTU, No.23 MU, MOTU, sold by Board of Trade to Stansted Fire School 17/07/68.

VP259 First flight 24/01/50. No.38 MU, Autolycus installation, loaned back to Avro (Woodford), converted to T Mk 4 (4/56), No.23 MU, MOTU, crashed Haldon Hill, Elgin 10/01/58, SOC 31/5/58.

VP260 First flight 22/1/51, No.38 MU, No.120 Sqn, No.23 MU, NEA, sold for scrap 29/09/63.

VP261 First flight 13/02/51. Woodford & A&AEE, No.38 MU, ASWDU, No.120 Sqn, crashed in sea near Berwick upon Tweed 25/06/52, SOC 11/08/52.

VP262 First flight 20/02/51. No.38 MU, No.120 Sqn, No.49 MU, MOTU, No.23 MU, NEA, sold for scrap 22/08/63.

VP263 First flight 17/03/51. RAE, A&AEE, No.38 MU, Nos. 42, 206 & 220 Sqns, Met. recce mods (3/58), Nos. 269 & 204 Sqns, MOTU, No.23 MU, NEA, sold for scrap 22/08/63.

VP264 First flight 08/03/51. CS(A), No.38 MU, No.236 OCU, Rolls-Royce Hucknall, No.38 MU, No.236 OCU/MOTU, No.23 MU, sold for scrap 22/11/62.

VP265 First flight 29/03/51. No.38 MU, No.120 Sqn, No.23 MU, No.49 MU, No.12 MU, No.23 MU, Nos.220, 206 & 269 Sqns, MOTU, No.23 MU, MOTU, No.23 MU, NEA, sold for scrap 23/10/63.

VP266 First flight 4/04/51. No.38 MU, No.120 Sqn, No.23 MU, Nos. 269 & 204 Sqns, MOTU, No.23 MU, NEA, sold for scrap 28/02/63.

VP267 First flight 13/04/51. No.38 MU, No.120 Sqn, No.23 MU, No.49 MU, No.23 MU, No.205 Sqn, withdrawn from service, reduced to components 1/12/62.

VP268 First flight 20/04/51. No.38 MU, No.236 OCU, MOTU, No.23 MU, NEA, sold for scrap 23/10/63.
VP281 to 294

VP281 First flight 24/4/51. No.38 MU, ASWDU, No.236 OCU/MOTU, No.23 MU, NAE, sold for scrap 28/02/63.

VP282 First flight 1/05/51. No.38 MU, ASWDU, A&AEE, No.49 MU, ASWDU, No.49 MU, Orange Harvest trials, No.23 MU, NAE, sold for scrap 31/05/62.

VP283 First flight 11/05/51. No.38 MU, No.224 Sqn, crashed 12/08/51, reduced to components & SOC 12/08/51.

VP284 First flight 23/05/51. No.38 MU, No.236 OCU, No.23 MU, No.204 Sqn, No.269 Sqn, No.23 MU, NAE, sold for scrap 28/02/63

VP285 First flight 26/05/51. No.38 MU, No.236 OCU, No.38 MU, A&AEE, Blue Silk trials, No.49 MU, TRE Defford, ASWDU, A&AEE, ASWDU, No.23 MU, MOTU, No.23 MU, NEA, sold for scrap 28/02/63.

VP286 First flight 31/05/51. No.38 MU, No.236 OCU, crashed at sea off Tarbet Ness, Cromarty 8/10/52, SOC 27/10/52

VP287 First flight 7/06/51. No.38 MU, No.224 Sqn, No.269 Sqn, No.240 Sqn, No.23 MU, NAE, sold as scrap 23/10/63.

VP288 First flight 15/06/51. Retained by Avro for sonobuoy trials, RAE, No.38 MU, ASWDU, Autolycus installation, No.220 Sqn, No.23 MU, No.205 Sqn, NEA, sold for scrap 5/8/64.

VP289 First flight 25/06/51. No.38 MU, No.224 Sqn, No.269 Sqn, No.23 MU, No.49 MU, No.206 Sqn, Met recce modifications, MOTU, No.23 MU, to Weston-super-Mare as instructional airframe 7730M, scrapped 4/66.

VP290 First flight 25/06/51. No.38 MU, No.224 Sqn, No.269 Sqn, No.23 MU, RAE West Freugh, No.23 MU, No.49 MU, No.38 MU, No.236 OCU/MOTU, No.23 MU, NEA, sold as scrap 22/11/62.

VP291 First flight 29/06/51. No.38 MU, No.224 Sqn, No.269 Sqn, No.23 MU, No.205 Sqn, No.23 MU, NEA, sold for scrap 5/08/64.

VP292 First flight 12/07/51. No.38 MU, No.236 OCU. No.23 MU, No.49 MU, No.23 MU, MOTU, No.23 MU, No.205 Sqn, FEAF MU Seletar, NEA, reduced to components, SOC 28/04/61.

VP293 First flight 18/07/51. No.38 MU, No.236 OCU, No.224 Sqn, No.23 MU, No.42 Sqn, No.206 Sqn, No.49 MU, CCMC, No.206 Sqn, converted to T Mk 4, No.23 MU, Phase 2 modifications, A&AEE, MOTU, RAE, sold to Strathallen Collection (5/76) and later scrapped by them, front fuselage preserved by the Avro Heritage Society at Woodford and on loan to the Newark Air Museum, extant.

VP294 First flight 18/07/51. No.38 MU, No.220 Sqn, No.236 OCU, No.224 Sqn, No.38 MU, RRE Blue Silk trials, No.23 MU, No.220 Sqn, No.206 Sqn, No.49 MU, No.269 Sqn, MOTU, No.205 Sqn, No.23 MU, No.205 Sqn, crash landed at Gan (5/62) reduced to components, SOC 18/05/62.

Specification Number: Air Ministry 42/46
Contract Number:
6/ACFT.3628/CB6(a), dated 18th May 1949
Shackleton MR Mk 1A
Quantity: 38, built as thirty-seven MR Mk 1As and one MR Mk 2 prototype WB818 to 837

WB818   First flight 1/08/51. No.38 MU, CAPMF modifications, No.269 Sqn, No.23 MU, No.205 Sqn, damaged at Gan during taxiing, stored at Seletar, reduced to components, SOC 28/04/62.

WB819   First flight 2/08/51. No.38 MU, No.224 Sqn, No.269 Sqn, No.38 MU, No.236 OCU, converted to T Mk 4, MOTU, Kinloss Station Flt, MOTU, SOC and burnt at Stanstead Fire School 14/06/68.

WB820   First flight 14/08/51. No.38 MU, No.224 Sqn, No.269 Sqn, No.23 MU, converted to T Mk 4, MOTU, reduced to components, SOC and put on the dump at RAF St. Mawgan (6/67).

WB821   First flight 17/08/51. No.38 MU, No.220 Sqn, No.236 OCU, CCMC modifications, No.236 OCU, No.206 Sqn, No.23 MU, MOTU, No.23 MU, NEA, sold as scrap 31/05/62.

WB822   First flight 17/08/51. No.38 MU, No.236 OCU, MOTU, No.23 MU, converted to T Mk 4, No.23 MU, MOTU, reduced to components, SOC, used for fire practice (8/68).

WB823   First flight 25/08/51. No.38 MU, No.220 Sqn, CCMC modifications, No.220 Sqn, No.206 Sqn, No.240 Sqn, No.23 MU, NEA, sold for scrap 29/05/63.

WB824   First flight 30/08/51. No.38 MU, No.220 Sqn, No.206 Sqn, No.236 OCU, MOTU, No.23 MU, NEA, sold for scrap 3/01/62.

WB825   First flight 31/08/51. No.38 MU, No.220 Sqn, No.23 MU, No.205 Sqn, stored at Seletar, reduced to components, SOC 8/08/61, sold for scrap.

WB826   First flight 3/09/51. No.38 MU, No.236 OCU, No.38 MU, No.206 Sqn, No.240 Sqn, No.269 Sqn, No.204 Sqn, MOTU, No.27 MU, NEA, sold for scrap 20/02/68.

WB827   First flight 12/09/51. No.38 MU, No.236 OCU/MOTU, No.23 MU, No.205 Sqn, No.23 MU, sold for scrap 5/08/64.

WB828   First flight 14/09/51. No.38 MU, No.220 Sqn, CAPMF modifications, No.204 Sqn, CAPMF modifications, No.240 Sqn, No.38 MU, No.240 Sqn, No.206 Sqn, No.204 Sqn, No.23 MU, NEA, sold for scrap 22/11/62.

WB829   First flight 21/09/51. No.38 MU, No.236 OCU, stored, No.236 OCU/MOTU, No.23 MU, FEAF Singapore (grounded on arrival), NEA, reduced to components, SOC 28/04/61.

WB830   First flight 26/09/51. No.38 MU, No.236 OCU, Rolls-Royce Hucknall (on loan), No.236 OCU/MOTU, No.23 MU, NEA, sold for scrap 31/05/62.

WB831   First flight 27/09/51. No.38 MU, No.220 Sqn, CCMC Autolycus installation, No.236 OCU, No.220 Sqn, converted to T Mk 4, MOTU, Phase 2 modifications, MOTU, crashed at St. Mawgan 17/05/67, reduced to components, SOC 7/06/67 used on fire dump.

WB832   First flight 3/10/51. No.38 MU, No.224 Sqn, No.206 Sqn, No.49 MU & CCMC modifications, No.206 Sqn, converted to T Mk 4, No.23 MU, MOTU, Phase 2 modifications, MOTU, to No.2 SofTT as instructional airframe 7885M, reduced to components and scrapped at Cosford.

WB833   Prototype MR Mk 2. First flight 17/06/52. A&AEE, RAE, Phase 1 modifications, A&AEE, Phase 2 modifications, A&AEE, RAE. A&AEE, ASWDU, Phase 3 modifications, Ballykelly Wing, crashed on the Mull of Kintyre 19/04/68, SOC 19/04/68.

WB834   First flight 11/10/51. No.38 MU, No.236 OCU/MOTU, No.23 MU, No.205 Sqn, stored at Selctar, reduced to components, SOC 8/08/61.

WB835   First flight 15/10/51. Trials with Mk 3 lifeboat, AIEU, A&AEE, returned to Avro, No.23 MU, No.240 Sqn, No.269 Sqn, No.23 MU, No.205 Sqn, No.23 MU, MOTU, No.23 MU, NEA, sold for scrap 23/10/63.

WB836   First flight 18/10/51. No.38 MU, No.224 Sqn, No.206 Sqn, No.220 Sqn, No.240 Sqn, No.23 MU,

No.205 Sqn, No.23 MU, NEA, sold for scrap 5/08/64.

WB837   First flight 24/10/51. No.38 MU, No.220 Sqn, converted to T Mk 4, No.23 MU, MOTU, No.27 MU, NEA, sold for scrap 3/02/69.

WB844 to 861

WB844   First flight 31/10/51. No.38 MU, No.224 Sqn, CAPMF modifications, No.224 Sqn, No.120 Sqn, converted to T Mk 4, No.23 MU, MOTU, Phase 2 modifications, MOTU, to No.2 SofTT as instructional airframe 8028M, scrapped at Cosford.

WB845   First flight 8/11/51. No.38 MU, No.224 Sqn, No.23 MU, No.236 OCU, No.23 MU, MOTU, No.23 MU, converted to T Mk 4, MOTU, No.27 MU, NEA, sold as scrap 12/03/69.

WB846   First flight 14/11/51. No.38 MU, No.224 Sqn, No.120 Sqn, MOTU, withdrawn from service, MOTU as instructional airframe 7561M, scrapped at Kinloss.

WB847   First flight 20/11/51. No.38 MU, No.236 OCU, converted to T Mk 4, No.23 MU, MOTU, Phase 2 modifications, MOTU, withdrawn from service, to instructional airframe 8020M, gate guardian RAF Kinloss, fire dump (3/69), SOC.

WB848   First flight 25/01/51. No.38 MU, No.236 OCU, MOTU, No.240 Sqn, No.23 MU, NEA, sold for scrap 23/10/63.

WB849   First flight 28/11/51. No.38 MU, JASS, No.120 Sqn, MOTU, No.23 MU, converted to T Mk 4, MOTU, withdrawn from service, to RAF Newton as instructional airframe 8026M (7/68), reduced to components and scrapped.

WB850   First flight 8/12/51. No.38 MU, JASS, No.240 Sqn, No.204 Sqn, No.23 MU, NEA, sold for scrap 29/05/63.

WB851   First flight 12/12/51. No.38 MU, JASS, No.269 Sqn, CCMC modifications, No.269 Sqn, No.38 MU, No.23 MU, No.220 Sqn, No.206 Sqn, No.269 Sqn, No.204 Sqn, ASWDU, No.23 MU, NEA, sold for scrap 28/02/63.

WB852   First flight 14/12/51. No.38 MU, No.224 Sqn, No.269 Sqn, Autolycus installation, No.269 Sqn, No.23 MU, NEA, sold for scrap 29/05/63.

WB853   First flight 19/12/51. No.38 MU, No.224 Sqn, CAPMF modifications, No.120 Sqn, CCMC inspection, MOTU, No.23 MU, NEA, sold for scrap 29/05/63.

WB854   First flight 29/12/51. No.38 MU, No.224 Sqn, No.120 Sqn, CCMC modifications, MOTU, conversion to T Mk 4 (cancelled), No.23 MU, No.205 Sqn, withdrawn from service, scrapped at Seletar.

WB855   First flight 2/01/52. No.38 MU, No.236 OCU, MOTU, No.269 Sqn, No.23 MU, NEA, sold for scrap 28/02/63.

WB856   First flight 9/01/52. No.38 MU, ASWDU, No.224 Sqn, CCCMC modifications, JASS, Autolycus installation, No.240 Sqn, No.49 MU, No.240 Sqn, No.204 Sqn, No.23 MU, NEA, sold for scrap 19/12/60.

WB857   First flight 16/01/52. No.38 MU, No.240 Sqn, CC(A) modifications, No.240 Sqn, No.49 MU, No.240 Sqn, No.23 MU, No.49 MU, No.240 Sqn, No.204 Sqn, No.269 Sqn, return to manufacturer for fatigue investigations, No.204 Sqn, No.27 MU, sold for scrap 31/05/62.

WB858   First flight 29/01/52. No.38 MU, No.240 Sqn, No.23 MU, No.236 OCU/MOTU, A&AEE, MOTU, converted to T Mk 4, MofA Air Fleet, A&AEE, No.23 MU, A&AEE, No.23 MU, MOTU, withdrawn from service, No.27 MU, NEA, sold for scrap 3/02/69.

WB859   First flight 30/01/52. No.38 MU, No.240 Sqn (renumbered No.203 Sqn 11/58), No.23 MU, sold for scrap 25/09/63.

WB860   First flight 7/02/52. No.38 MU, No.240 Sqn, No.49

MU, No.269 Sqn, No.203 Sqn, No.204 Sqn, No.23 MU. NEA, sold for scrap 28/02/63.

WB861   First flight 20/02/52. No.38 MU, No.240 Sqn, No.49 MU, No.240 Sqn, crash landed 5/09/57, reduced to components, SOC 6/09/57.

Specification Number:
Air Ministry 42/46
Contract Number:
6/ACFT.5047/CB6(a), dated August 1950
Shackleton MR Mk 1A
Quantity: 20, built as ten MR Mk 1As and ten MR Mk 2s
WG507 to 511

WG507   First flight 26/02/52. No.38 MU, No.240 Sqn, CCMC modifications, No.240 Sqn, No.49 MU, No.240 Sqn, No.23 MU, NEA, scrapped 31/05/62.

WG508   First flight 8/03/52. No.38 MU, No.240 Sqn, No.206 Sqn, No.220 Sqn, No.206 Sqn, No.23 MU, NEA, sold for scrap 29/05/63.

WG509   First flight 13/03/52. No.38 MU, No.240 Sqn, No.49 MU, No.240 Sqn, withdrawn from service, No.23 MU, sold for scrap 25/09/63.

WG510   First flight 30/03/52. No.38 MU, No.42 Sqn, No.206 Sqn, CCMC Autolycus installation, No.206 Sqn, No.23 MU, NEA, sold for scrap 28/02/63

WG511   First flight 7/04/52. No.38 MU, No.42 Sqn, No.120 Sqn, No.49 MU, Autolycus installation, No.23 MU, converted to T Mk 4, No.23 MU, MOTU, Phase 2 modifications, MOTU, withdrawn from service, reduced to components, SOC 3/08/66, to No.71 MU and front fuselage cut off, RAF Museum, sold to Cornwall Aero Park, extant.
WG525 to 529

WG525   First flight 18/04/52. No.38 MU, No.42 Sqn, No.220 Sqn, No.23 MU, No.205 Sqn, landing accident at Gan, returned to UK, No.205 Sqn, No.23 MU, NEA, sold for scrap 5/08/64.

WG526   First flight 6/05/52. No.38 MU, No.42 Sqn, No.206 Sqn, CCMC Autolycus installation, No.206 Sqn, No.220 Sqn, No.23 MU, MOTU, ASWDU, SOC, reduced to components and scrapped 11/07/61.

WG527   First flight 15/05/52. No.38 MU, No.42 Sqn, converted to T Mk 4, No.23 MU, MOTU, Phase 2 modifications, MOTU, withdrawn from service, No.27 MU, NEA, sold for scrap 12/03/69.

WG528   First flight 18/05/52. No.38 MU, No.42 Sqn, No.206 Sqn, No.23 MU, NEA, sold for scrap 29/05/63.

WG529   First flight 24/06/52. No.38 MU, No.42 Sqn, No.206 Sqn, No.240 Sqn, No.23 MU, NEA, sold for scrap 25/11/63.

Specification Number:
Air Ministry 42/46
Contract Number:
6/ACFT.5047/CB6(a), dated August 1950
Shackleton MR Mk 2
Quantity: 20, built as ten MR Mk 1As
(See WG507-WG529) and ten MR Mk 2 WG530 to 533

WG530   First flight 15/08/52. A&AEE, No.23 MU, No.49 MU, No.120 Sqn, No.49 MU (Phase 1 Mods), No.120 Sqn, No.49 MU, No.224 Sqn, No.49 MU, No.42 Sqn, Phase 2 mods, No.205 Sqn, No.27 MU, NEA sold for scrap 3/09/68.

WG531   First flight 21/08/52. CS(A) trials, No.38 MU (cancelled), RAF St. Eval (trials), St. Eval Station Flight, No.42 Sqn, CCMC mods, believed to have been in mid-air collision with WL743 (missing) 11/01/55.

WG532   First flight 12/09/52. CS(A) for Glow Worm trials, No.23 MU, ASWDU, A&AEE for Glow Worm trials, No.42 Sqn, CCMC, No.42 Sqn, No.120 Sqn, No.49 MU (Phase 1 mods), No.224 Sqn, Phase 2 mods,

No.224 Sqn, No.205 Sqn, No.27 MU, NEA sold for scrap 03/09/68.

WG533   First flight 18/09/52. RAF Handling Squadron (Manby), No.23 MU, No.42 Sqn, No.49 MU, No.42 Sqn, ASWDU, Phase 1 mods, No.38 Sqn, Phase 2 mods, No.224 Sqn, No.38 Sqn, converted to T Mk 2, MOTU/No.236 OCU, No.5 MU, No.225 Sqn (after Phase 3 mods completed), No.204 Sqn, No.32 MU, NEA for sale as scrap (cancelled), RAF St Athan dump (1974).
WG553 to 558

WG553   First flight 2/10/52. No.23 MU, ASDWU, CAPMF mod, RAE, ASWDU, No.49 MU (Phase 1 mods), ASWDU, Phase 2 mods, No.205 Sqn, No.27 MU, NEA sold for scrap 24/06/68.

WG554   First flight 10/10/52. No.23 MU, No.42 Sqn, Phase 1 mods, No.224 Sqn, Phase 2 mods, ASWDU, No.60 MU (repair), ASWDU, No.205 Sqn, converted to T Mk 2, MOTU, No.71 MU (repaired), MOTU, No.5 MU, Phase 3 mods, St Mawgan Gulf Detachment, No.210 Sqn, No.32 MU, NEA, St Athan dump.

WG555   First flight 21/10/52. No.23 MU, No.42 Sqn, Phase 1 mods, No.210 Sqn, Phase 2 mods, No.204 Sqn, No.23 MU (repaired), No.204 Sqn, Phase 3 mods, No.204 Sqn, Majunda Detachment Support Unit, No.204 Sqn, SOC, RAF Fire Fighting School RAF Catterick (09/05/72).

WG556   First flight 28/10/52. No.23 MU, No.42 Sqn, No.23 MU, No.5 MU, No.120 Sqn, No.37 Sqn, Phase 1 mods, No.224 Sqn, MofA Air Fleet, Phase 2 mods, A&AEE, RAE for Jezebel trials. ASWDU for Jezebel trials, Ballykellly Wing, No.32 MU, Battle Damage Repair Flight RAF Lossiemouth, Fire Section RAF Lossiemouth as instructional airframe No.8651M, scrapped (July 1982).

WG557   First flight 5/11/52. No.23 MU, No.206 Sqn, No.220 Sqn, No.228 Sqn, No.38 MU, No.23 MU, No.5 MU, No.23 MU, RAE Armament Flight, ETPS, RAE Armament Flight, SOC (28/10/64), RAE Farnborough dump.

WG558   First flight 11/11/52. No.23 MU, No.206 Sqn, CAPMF mods, No.224 Sqn, Phase 1 mods, No.42 Sqn, No.204 Sqn, A&AEE, Phase 2 mods, No.210 Sqn, No.71 MU (repairs), No.210 Sqn, converted to T Mk 2, MOTU, No.71 MU (repairs), MOTU, No.71 MU (repairs), MOTU, No.5 MU, Phase 3 mods, RAF St. Mawgan Gulf Detachment, No.210 Sqn, No.32 MU, NEA, reduced to components 25/01/74.

Specification Number:
Air Ministry 42/46
Contract Number:
6/ACFT.6129/CB6(a), dated December 1950
Shackleton MR Mk 2A
Quantity: 20 (conversions)

WL737   First flight 17/11/52, No.23 MU, CS(A), No.23 MU, No.220 Sqn, CAPMF mods, No.42 Sqn, Phase 1 mods, No.42 Sqn, Phase 2 mods, MofA (glide path aerial trials), Phase 3 mods, RAE Bedford, A&AEE, returned to Avro at Woodford, A&AEE, No.210 Sqn, No.205 Sqn, No.32 MU, NEA, reduced to components 31/08/73.

WL738   First flight 25/11/52. No.23 MU, No.240 Sqn, CAPMF mods, No.240 Sqn, No.37 Sqn, No.49 MU, Phase 1 mods, No.37 Sqn, Phase 2 mods, No.37 Sqn, Ministry of Technology at A&AEE for radio altimeter trials, Ballykelly Wing, Bitteswell for mods, Ballykelly Wing, No.204 Sqn, No.5 MU, No.204 Sqn, No.5 MU, No.8 Sqn, SOC 14/10/77 (cancelled), instructional airframe No.8567M as gate guard at RAF Lossiemouth, scrapped in 1991.

WL739   First flight 9/12/54. CAPMF mods, No.224 Sqn, Phase 1 mods, No.42 Sqn, No.204 Sqn, A&AEE, Phase 2 mods, No.210 Sqn, No.71 MU (repairs),

No.210 Sqn, converted to T Mk 2, MOTU, Bitteswell for mods, No.71 MU (repairs), No.5 MU, Phase 3 mods, RAF St. Mawgan Gulf Detachment, No.210 Sqn, RAF Fire Fighting School RAF Catterick (11/71).

WL740   First flight 18/12/52. CS(A), No.23 MU, No.204 Sqn, CAPMF mods, No.204 Sqn, A&AEE, No.38 Sqn, Safi (Malta) for storage, Phase 1 & 2 mods, No.38 Sqn, converted to T Mk 2 but not completed, reduced to components and SOC 28/02/68.

WL741   First flight 18/12/52. No.23 MU, No.224 Sqn, CAPMF mods, Phase 1 mods, No.42 Sqn, No.49 MU (repair), No.42 Sqn, No.224 Sqn, Phase 2 mods, No.205 Sqn, Phase III mods, No.205 Sqn, No.5 MU, converted to AEW Mk 2, No.8 Sqn (named P C Knapweed), No.60 MU for mods, No.8 Sqn, withdrawn from service, CTE Manston as instructional airframe No.8692M, burnt on site (June 1982).

WL742   First flight 23.12.52. No.23 MU, No.206 Sqn, No.42 Sqn, Phase 1 mods, No.224 Sqn, Phase 2 mods, No.203 Sqn, No.60 MU (repairs - twice), No.204 Sqn, Ballykelly Wing, No.27 MU, NEA sold for scrap 26/06/68.

WL743   First flight 20/01/53. No.23 MU, No.220 Sqn, No.42 Sqn, believed to have been involved in mid-air collision with WG531 (missing) 11/01/55.

WL744   First flight 15/01/52. No.23 MU, No.120 Sqn, CAPMF mods, No.228 Sqn, No.38 MU, No.23 MU, No.5 MU, No.23 MU, No.204 Sqn, No.42 Sqn, No.49 MU, Phase 1 mods, No.38 MU, Phase 2 mods, No.37 Sqn, Ballykelly Station Flight, SOC 1/11/66 (on dump at Ballykelly, subsequently scrapped).

WL745   First flight 22/01/53. No.23 MU. No.220 Sqn, No.42 Sqn, No.120 Sqn, Phase 1 mods, No.204 Sqn, Phase 2 mods, No.205 Sqn, Phase 3 mods, Ballykelly Wing, Woodford (for performance trials), MofA Air Fleet, converted to AEW Mk 2, A&AEE, Bitteswell (update to production AEW Mk 2 standards), No.8 Sqn (named Sage), withdrawn from service (6/81), RAF Fire Fighting School RAF Catterick as instructional airframe No.8698M, subsequently burnt there during 1983.

WL746   First flight 28/01/53. No.23 MU, No.269 Sqn, crashed into sea off Argyll 11/12/53, salvaged and SOC 12/12/53.

WL747   First flight 5/02/53. No.23 MU, No.269 Sqn, No.204 Sqn, No.42 Sqn, Phase 1 mods, No.210 Sqn, Phase 2 mods, No.37 Sqn, Phase 3 mods, No.210 Sqn, Ballykelly Wing, No.204 Sqn, converted to AEW Mk 2, No.5 MU, No.8 Sqn (named Florence), Bitteswell (refurbished), No.8 Sqn, withdrawn from service (7/91) sold to Savvas Constantinides, extant at Paphos Airport Cyprus.

WL748   First flight 6/02/53. No.23 MU, No.240 Sqn, No.204 Sqn, No.269 Sqn, No.210 Sqn, No.204 Sqn, Phase I mods, No.210 Sqn, Phase 2 mods, No.210 Sqn, Phase 3 mods, RRE Pershore for infrared scanner trials, No.205 Sqn, Ballykelly Wing, RRE (infrared equipment removed), No.204 Sqn, Majunda Detachment Support Unit, No.204 Sqn, RAF Fire Fighting School RAF Catterick (8/05/72) subsequently burnt.

WL749   First flight 18/02/53. No.23 MU, No.120 Sqn, damaged and reduced to components 14/05/53.

WL750   First flight 23/02/53. No.23 MU, No.269 Sqn, CAPMF mods, No.204 Sqn, Autolycus installation, No.120 Sqn, No.269 Sqn, Phase 1 mods, No.224 Sqn, Phase 2 mods, No.203 Sqn, No.60 MU (repairs), No.203 Sqn, No.204 Sqn, converted to T Mk 2, MOTU, No.5 MU, Phase 3 mods, No.205 Sqn, RAF Fire Fighting School RAF Catterick (cancelled), CTE Manston, SOC (1971?).

WL751   First flight 5/03/52. No.23 MU, No.224 Sqn,

CAPMF mods, No.49 MU, No.224 Sqn, Phase 1 mods, No.204 Sqn, Phase 2 mods, No.210 Sqn, No.23 MU (repairs), Phase 3 mods, HSA for stall warning system trials, Ballykelly Wing, Bitteswell for mods, No.5 MU, No.204 Sqn, Majunda Detachment Support Unit, No.204 Sqn, sold to Shackleton Aviation Baginton (4/05/72), scrapped January 1975.

WL752   First flight 7/03/53. No.23 MU, No.120 Sqn, No.23 MU, No.224 Sqn, Phase 1 mods, No.37 Sqn, Phase 2 mods, RAF Handling School (Manby), No.23 MU, No.37 Sqn, No.27 MU, NEA, sold for scrap 7/10/68.

WL753   First flight 16/03/53. No.23 MU, No.224 Sqn, CAPMF mods, Phase 1 mods, RAF Eastleigh Station Flight (Kenya) & storage, No.37 Sqn, Phase 2 mods, No.203 Sqn, No.204 Sqn, Ballykelly Wing, No.27 MU, NEA, sold for scrap 12/03/69.

WL754   First flight 18/03/53. No.23 MU, Overseas Ferry Unit (Benson), No.137 MU, No.37 Sqn, No.137 MU (repairs), No.37 Sqn, CAPMF mods, No.137 MU (major servicing), No.38 Sqn, No.137 MU (repairs), No.37 Sqn, No.38 Sqn, No.37 Sqn, Phase 1 mods, Phase 2 mods, No.42 Sqn, Phase 3 mods, No.205 Sqn, Bitteswell (fitment of stall warning system), Ministry of Technology trials at Woodford, Ballykelly Wing, No.204 Sqn, No.5 MU, converted to AEW Mk 2, No.5 MU, No.8 Sqn (named Paul), withdrawn from service (January 1981), RAF Valley for fire practice and crash rescue as Instructional Airframe No.8865M (22/01/81).

WL755   First flight 30/03/52. No.23 MU, Overseas Ferry Unit (Benson), No.137 MU, No.37 Sqn, CAPMF mods, No.137 MU (major servicing), No.37 Sqn, No.49 MU (Autolycus installed), No.137 MU, No.38 Sqn, Phase 1 mods, No.38 Sqn, Phase 2 mods, No.224 Sqn, Phase 3 mods, No.60 MU (repairs), No.204 Sqn, Majunda Detachment Support Unit, No.204 Sqn, RAF Fire Fighting School Catterick, subsequently burnt there (December 1977).

WL756   First flight 1/04/53. No.23 MU, Overseas Ferry Unit (Benson), No.37 Sqn, CAPMF mods, No.137 MU (repairs), No.38 Sqn, No.49 MU, No.103 MU (repairs), Phase 1 mods, No.37 Sqn, Phase 2 mods, No.205 Sqn, Phase 3 mods, Ballykelly Wing, No.204 Sqn, No.5 MU, converted to AEW Mk 2, No.5 MU, No.8 Sqn (named Mr Rusty), withdrawn from service (1/7/91), to RAF St. Mawgan for crash rescue training, subsequently burnt there by 1998.

WL757   First flight 10/04/53. No.23 MU, Overseas Ferry Unit (Benson), No.37 Sqn, CAPMF mods, No.137 MU, No.37 Sqn, No.137 MU, No.38 Sqn, Phase 1 mods, No.210 Sqn, Phase 2 mods, No.224 Sqn, Phase 3 mods, No.205 Sqn, No.5 MU, converted to AEW Mk 2, No.5 MU, No.8 Sqn (named Brian), withdrawn from service (July 1991), sold to Savvas Constantinides (3/7/91), extant Paphos Airport (Cyprus).

WL758   First flight 17/04/53. No.23 MU, No.120 Sqn, CAPMF, No.224 Sqn, No.49 MU, Phase 1 mods, Phase 2 mods, No.38 Sqn, Phase 3 mods, Ballykelly Wing, No.204 Sqn, RAF Fire Fighting School Catterick, subsequently burnt there by 1975.

WL759   First flight 24/04/53. No.23 MU, No.37 Sqn, CAPMF mods, No.137 MU (repairs), No.37 Sqn, No.137 MU (repairs), No.37 Sqn, No.49 MU, No.204 Sqn, Phase 1 mods, No.38 Sqn, Phase 2 mods, to MofA on loan for Mk 44 torpedo, new intercom system and revised tailwheel trials, A&AEE (Mk 44 torpedo and marine marker trials), ASWDU, No.205 Sqn, SOC (11/11/68), broken up during 1969.

WL785   First flight 10/05/53. No.23 MU, No.37 Sqn, CAPMF mods, No.137 MU (repairs), No.37 Sqn, No.38 Sqn, Phase 1 mods, No.42 Sqn, Phase 2 mods, No.42 Sqn, MofA Air Fleet, A&AEE (sonobuoy trials), Phase 3 mods, No.204 Sqn, Ballykelly Wing, Bitteswell for mods, No.204 Sqn, No.32 MU, CTE Manston, SOC and reduced to components 30/06/71.

WL786 First flight 12 /05/53. No.23 MU, No.137 MU, No.37 Sqn, CAPMF mods, No.137 MU, No.37 Sqn, No.137 MU, No.37 Sqn, Phase 1 mods, No.38 Sqn, Phase II mods, No.205 Sqn, crashed in the India Ocean 5/11/67, SOC 6/11/67.

WL787 First flight 18/05/53. No.23 MU, No.137 MU, No.38 Sqn, CAPMF mods, No.38 Sqn, No.49 MU (Autolycus installed & mods), Phase 1 mods, No.137 MU, No.38 Sqn, No.37 Sqn, Phase 2 mods, No.210 Sqn, No.204 Sqn, converted to T Mk 2, MOTU, No.5 MU, Phase 3 mods, RAF St. Mawgan Gulf Detachment, No.210 Sqn, RAF Fire Fighting School Catterick (cancelled), No.8 Sqn (for crew training, named Mr McHenry), damaged & repaired then renamed Dylan, withdrawn from service and used for fire fighting practice (3/01/74), broken up by March 1974.

WL788 First flight 4/06/53. No.23 MU, No.137 MU, No.37 Sqn, CAPMF mods, No.137 MU, No.49 MU (Autolycos installed), No.37 Sqn, No.137 MU (repairs), No.37 Sqn, No.49 MU, No.38 Sqn, No.137 MU, No.38 Sqn, Phase 1 mods, No.38 Sqn, No.137 MU (repairs), Phase 2 mods, No.42 Sqn, No.210 Sqn, No.37 Sqn, No.27 MU, NEA and SOC, sold for scrap 28/03/69.

WL789 First flight 10/06/53. No.38 MU, MAD tail boom installation, CAPMF mods, ASWDU (for MAD trials), CAPMF mods, ASWDU, MAD boom removed (4/58), No.49 MU (Phase 1 mods), No.224 Sqn, Phase 2 mods, No.224 Sqn, No.38 Sqn, No.205 Sqn, No.27 MU, NEA and broken up, sold for scrap 28/03/69.

WL790 First flight 23/06/53. No.38 MU, No.240 Sqn, No.204 Sqn, No.49 MU (Autolycus installed), No.204 Sqn, No.38 MU (repairs), No.204 Sqn, No.269 Sqn, No.210 Sqn, No.49 MU (Phase 1 mods), No.210 Sqn, Phase 2 mods, No.205 Sqn, Phase 3 mods, No.205 Sqn, No.5 MU, converted to AEW Mk 2, No.5 MU, No.8 Sqn (named Mr McHenry and later renamed Zebedee), Bitteswell for re-sparring, No.8 Sqn, withdrawn from service (1/07/91), sold to Air Atlantique, ownership passed to Shackleton Preservation Trust, sold to Polar Aviation Museum, Minnesota and delivered as N790WL 7/09/94. extant and airworthy.

WL791 First flight 16/06/52. No.23 MU, No.38 MU, No.137 MU, No.37 Sqn, No.49 MU (Autolycus installed), No.38 Sqn, Phase 1 mods, No.38 Sqn, Phase 2 mods, Ballykelly Wing, No.27 MU, NEA, broken up, sold for scrap 28/03/69.

WL792 First flight 1/07/53. No.38 MU, No.204 Sqn, No.49 MU (Autolycus installed), No.204 Sqn, No.38 Mu, No.5 MU, No.224 Sqn, crashed during air display on Gibraltar 14/09/57. reduced to components 11/11/57.

WL793 First flight 15/07/53. No.23 MU, No.38 Sqn, No.137 MU, No.38 Sqn, No.137 MU, Phase 1 mods, No.204 Sqn, No.Phase 2 mods, No.210 Sqn, Phase 3 mods, No.210 Sqn, Ballykelly Wing, Bitteswell for mods, No.5 MU, converted to AEW Mk 2, No. Sqn (named Ermintrude), MofA Air Fleet, A&AEE, No.8 Sqn, Bitteswell for re-sparring, No.8 Sqn, withdrawn from service, to Battle Damage Repair Fleet at Lossiemouth as Instructional Airframe No.8675M, scrapped July 1982.

WL794 First flight 7/08/53. No.23 MU, No.38 Sqn, crashed into Mediterranean off Gozo 12/02/54, posted as missing and SOC.

WL795 First flight 17/08/53. No.23 MU, CAPMF mods, No.204 Sqn, No.49 Mu(Autolycus installed),

No.269 Sqn, No.210 Sqn, No.204 Sqn, Phase 1 and 2 mods, No.38 Sqn, Phase 3 mods, No.205 Sqn, No.5 MU, converted to AEW Mk 2, No.5 MU, No.8 Sqn (named Rosalie), withdrawn from service, RAF St. Mawgan for fire and rescue training as Instructional Airframe No.8753M 24/11/81 (cancelled) stored then refurbished and put on display at St Mawgan. extant.

WL796 First flight 23/08/53. No.23 MU, No.38 Sqn, No.37 Sqn, Phase 1 & 2 mods, No.204 Sqn, No.71 MU (repairs), No.205 Sqn, No.27 MU, NEA, sold for scrap 7/10/68.

WL797 First flight 15/09/53. No.23 MU, No.38 Sqn, No.137 MU (mods), No.37 Sqn, No.49 MU (Autolycus installed), No.37 Sqn, No.38 Sqn, No.137 MU (repairs), No.27 MU (storage), No.23 MU, RAF North Front Gibraltar, No.224 Sqn, No.49 MU (Phase 1 mods)No.204 Sqn, No.210 Sqn, Phase 2 mods, No.42 Sqn, No.37 Sqn, No.27 MU, NEA, sold for scrap 7/10/68.

WL798 First flight 17/09/53. No.23 MU, No.38 Sqn, No.137 MU (repairs), No.38 Sqn, No.49 MU (Autolycus installed), No.137 MU, No.38 Sqn, No.137 MU (repairs), Phase 1 mods, No.38 Sqn, Phase 2 mods, No.205 Sqn, Ballykelly Wing, No.2 SofTT Cosford as Instructional Airframe No.8114M, to RAF Lossiemouth to act as spares source for remaining aircraft flown by No.8 Sqn, front fuselage sold to private collector in Elgin, extant.

WL799 First flight 18/09/53. No.23 MU, No.38 Sqn, No.49 MU (Autolycus installed), destroyed in a hangar fire at Langar 22/12/55.

WL800 First flight 1/10/53. No.23 MU, No.137 MU, No.38 Sqn, No.49 MU (Autolycus installed), Langar (for mods & service), No.38 Sqn, No.37 Sqn, Phase 1 mods, No.42 Sqn, No.224 Sqn, Phase 2 mods, ASWDU for trials, No.203 Sqn, No.23 MU (repairs), No.203 Sqn, Phase 3 mods, No.204 Sqn, Ballykelly Wing, Bitteswell for mods, Ballykelly Wing, No.5 MU, No.204 Sqn, Majunga Detachment Support Unit, No.204 Sqn, RAF Fire Fighting School Catterick, SOC & burnt 19/05/72.

WL801 First flight 10/10/53. No.23 MU, No.137 MU, No.38 Sqn, No.49 MU (Autolycus installed), No.38 Sqn, Langar for mods, No.38 Sqn, Phase 1 mods, No.38 Sqn, Phase 2 mods, No.42 Sqn, Phase 3 mods, ASWDU, No.5 MU, No.8 Sqn for crew training, withdrawn from service, to the Aerospace Museum Cosford and subsequently scrapped in 1991.

Specification Number:
Air Ministry 42/46
Contract Number:
6/ACFT.6408/CB6(a), dated 8th February 1951
Shackleton MR Mk 2
Quantity: 40, built as nineteen MR Mk 2s (WR951 - WR969) and twenty-one MR Mk 3s (See WR970 to WR990)

WR951 First flight 20/10/53. No.38 MU, No.204 Sqn, CAPMF mods, No.228 Sqn, No.42 Sqn, No.49 MU (Autolycus installed), Phase 1 mods, No.204 Sqn, Phase 2 mods, No.224 Sqn, No.204 Sqn, Ballykelly Wing, Phase 3 mods, planned for conversion to T Mk 2 but cancelled, SOC and reduced to components 28/02/69.

WR952 First flight 27/10/53. No.38 MU, No.206 Sqn, No.42 Sqn, No.49 MU (Autolycus installed), Phase 1 mods, No.42 Sqn, Phase 2 mods, No.42 Sqn, No.205 Sqn, No.204 Sqn, RAF St. Athan, NEA, broken up for scrap 26/09/73.

WR953 First flight 6/11/53. No.38 MU, No.228 Sqn, No.224 Sqn, No.49 MU, No.224 Sqn, British Forces Arabian Peninsula, No.224 Sqn, No.49 MU (Phase 1 mods), No.42 Sqn, Langar for oxygen system trials, Phase 2 mods, No.224 Sqn, No.42 Sqn, No.205 Sqn, No.27 MU, NEA, Fire Fighting Training Kinloss, SOC as scrap during 1968.

WR954　First flight 19/11/53. No.23 MU, No.37 Sqn, No.49 MU (Autolycus installed(, No.37 Sqn, No.137 MU (mods), No.37 Sqn, Langar for major service, No.37 Sqn, No.38 Sqn, Phase 1 mods, No.210 Sqn, No.37 Sqn, Phase 2 mods, No.205 Sqn, Phase 3 mods, No.205 Sqn, RAF St. Athan, NEA, SOC as scrap 9/05/73.

WR955　First flight 27/11/53. No.38 MU, No.120 Sqn, No.224 Sqn, No.120 Sqn, No.210 Sqn, No.204 Sqn, Phase 1 mods, No.42 Sqn, Phase 2 mods, No.42 Sqn, Phase 3 mods, Ballykelly Wing, No.204 Sqn, A&AEE for landing performance trials on AEW.2 programme, Fire Fighting Training at RAF Brize Norton, SOC 1971.

WR956　First flight 10/12/53. No.23 MU, CAPMF mods, No.228 Sqn, No.49 MU (Autolycus installed), No.120 Sqn, No.269 Sqn, No.210 Sqn, Phase 1 & 2 mods, No.38 Sqn, Phase 3 mods, Ballykelly Wing, crash-landed at Ballykelly 1/04/68. SOC and put on the dump at Ballykelly 1/04/68.

WR957　First flight 21/12/53. No.38 MU, CAPMF mods, No.228 Sqn, No.49 MU (Autolycus mods), No.228 Sqn, Phase 1 mods, No.204 Sqn, No.224 Sqn, No.203 Sqn, No.204 Sqn, No.27 MU, NEA, sold for scrap 26/06/68.

WR958　First flight 8/01/54. No.23 MU, CAPMF mods, No.228 Sqn, No.49 MU (Autolycus installed), No.228 Sqn, No.42 Sqn, Phase 1 mods, No.42 Sqn, Phase 2 mods, No.42 Sqn, No.71 MU (repairs), Ballykelly Wing, No.27 MU, NEA, sold for scrap 3/09/68.

WR959　First flight 20/01/54. No.38 MU, CAPMF mods, No.228 Sqn, No.49 MU (Autolycus installed), No.228 Sqn, No.42 Sqn, Phase 1 mods, No.37 Sqn, ASF Eastleigh (Kenya), Phase 2 mods, No.203 Sqn, No.23 MU, No.42 Sqn, No.203 Sqn, No.205 Sqn, SOC and reduced to components at Changi 3/09/68.

WR960　First flight 5/02/54. No.23 MU, CAPMF mods, No.228 Sqn, No.49 MU (Autolycus installed), No.228 Sqn, No.49 MU, No.42 Sqn, Phase 1 mods, No.49 MU, No.210 Sqn, Phase 2 mods, MofA Air Fleet, A&AEE Phase 3 armament trials, Langar (service), No.205 Sqn, No.5 MU, converted to AEW Mk 2, No.5 MU, No.8 Sqn (named Dougal), Bitteswell for re-sparring, No.8 Sqn, withdrawn from service, flew into Cosford as Instructional Airframe No.8772M, transported to the Museum of Science and Industry in Manchester and re-erected for display 27/01/83, extant.

WR961　First flight 12/02/54. No.38 MU, CAPMF mods, No.228 Sqn, No.49 MU (Autolycus installed), No.228 Sqn, No.224 Sqn, No.204 Sqn, No.37 Sqn, Phase 1 & 2 mods, No.38 Sqn, Phase 3 mods, Ballykelly Wing, Bitteswell for mods, Ballykelly Wing, Bitteswell for mods, No.204 Sqn, Majunda Detachment Support Unit, No.5 MU, NEA, sold for scrap 1/02/78.

WR962　First flight 2/03/54. No.23 MU, No.228 Sqn, No.49 MU (Autolycus installed), No.228 Sqn, No.220 Sqn, No.228 Sqn, C(A), A&AEE for armament and Lindholme gear trials, No.204 Sqn, British Forces Arabian Peninsula, No.37 Sqn, Phase 3 mods, No.37 sqn, No.27 MU, NEA, sold for scrap 28/03/69.

WR963　This aircraft is owned by Air Atlantique with the radome removed to look like an MR Mk 2, extant.

WR964　First flight 18/03/54. Retained at Woodford, A&AEE for tropical cooling trials, Khartoum for tropical trials, A&AEE, Woodford for manufacturer's trials, No.38 MU, No.49 MU (Autolycus installed), No.38 MU, No.38 Sqn, No.49 MU, No.38 Sqn, Phase 1 & 2

mods, No.204 Sqn, No.23 MU (repairs), No.204 Sqn, converted to T Mk 2, MOTU, No.71 MU (repairs), No.32 MM, NEA, sold as scrap 15/11/71.

WR965　First flight 7/04/54. No.23 MU, No.37 Sqn, No.49 MU (Autolycus installed), No.37 Sqn, Langar for major service, No.37 Sqn, No.38 Sqn, No.49 MU, Phase 1 mods, No.224 Sqn, Phase 2 mods, A&AEE flame float launching system trials, No.203 Sqn, No.23 MU (repairs), No.203 Sqn, No.60 MU (repairs), No.203 Sqn, Phase 3 mods, Ballykelly Wing, Bitteswell (new tailplane installed), Ballykelly Wing, No.204 Sqn, converted to AEW Mk 2, No.5 MU, No.89 Sqn (named Dill and later renamed Rosalie), Bitteswell for re-sparring, No.8 Sqn, crashed in the Outer Hebrides 30/4/90, SOC 30/4/90.

WR966　First flight 28./04/54. No.38 MU, Ballykelly, No.49 MU (Autolycus installed), JASS Flight, No.220 Sqn, No.228 Sqn, No.37 Sqn, Phase 1 & 2 mods, No.204 Sqn, No.23 MU (repairs), No.210 Sqn, converted to T Mk 2, MOTU/No.236 OCU, No.5 MU (Phase 3 mods), No.204 Sqn, No.32 MU, NEA reduced to components and scrapped 22/06/73.

WR967　First flight 17/05/54. No.23 MU, JASS Flight, No.49 MU(Autolycus installed), JASS Flight, No.42 Sqn, No.49 MU (Phase 1 mods), No.42 Sqn, Phase 2 mods, No.38 Sqn, No.205 Sqn (on loan), No.38 Sqn, converted to T Mk 2, Ballykelly Wing, Bitteswell (for mods), Ballykelly Wing, MOTU, No.71 MU (repairs), MOTU, No.71 MU (repairs), MOTU, No.5 MU, No.210 Sqn, RAF Fire Fighting School (cancelled), No.8 Sqn for crew training (named Zebedee), damaged 7/09/72, wings removed and fuselage converted to AEW Mk 2 training simulator, No.8 Sqn as Instructional Airframe No.8398M (named Dodo), scrapped in 1991.

WR968　First flight 17/06/54. No.38 MU, No.224 Sqn, CCMC(Autolycus installed), No.224 Sqn, No.28 MU (repairs), No.120 Sqn, No.49 MU (Phase 1 mods), C(A) on loan for IFF Mk 10 trials, A&AEE for IFF Mk 10 clearance trials, No.224 Sqn, No.37 Sqn (temporary), No.224 Sqn, No.49 MU, Langar (Phase 2 mods), Ballykelly Wing, crash-landed and burnt out at Ballykelly 20/10/61, SOC 20/10/61.

WR969　First flight 10/5/54. No.23 MU, JASS Flight, No.49 MU (Autolycus installed), No.220 Sqn, No.228 Sqn, No.224 Sqn, No.49 MU (Phase 1 mods)No.210 Sqn, Phase 2 mods, No.204 Sqn, No.224 Sqn, No.38 Sqn, converted to T Mk 2, MOTU, No.5 MU, Phase 3 mods, No.205 Sqn, NEA, SOC as scrap 22/7/74.

Specification Number:
　Air Ministry 42/46

Contract Number:
　6/ACFT.6408/CB6(a), dated 8th February 1951

Shackleton MR Mk 2
　Quantity: 40, built as nineteen MR Mk 2s
　(See WR951 - WR969) and twenty-one MR Mk 3s
　(WR970 to WR990)

WR970　First flight 2/09/55. C(A), Woodford for manufacturer's trials, A&AEE for handling trials, MofS (not cleared for service), Woodford (stall warning trials), crashed at Foolow Derbyshire 7/12/56, SOC 28/11/57.

WR971　First flight 28/05/56. Woodford for manufacturer's trials, C(A), A&AEE for armament trials, Woodford to be updated to production standard, Phase 1 mods, No.120 Sqn, Phase 2 mods, No.120 Sqn, Phase 3 mods, No.201 Sqn, Langar for Viper installation, Kinloss Wing, No.60 MU, Bitteswell for mods, No.32 MU, No.2 SofTT as Instructional Airframe No.8119M, fuselage only sold to Wellesley Aviation (Narborough) in 1988, displayed at Fenland and West Norfolk Aviation Museum, extant.

WR972   First flight 6/11/56, C(A), A&AEE for radar, radio, navigation and photographic trials, Woodford for Autolycus trials, A&AEE official release trials, Woodford for radio mod trials, purchased by MofA for RAE, A&AEE for sonobuoy and mixed bomb clearance trials, Langar for aerial vibration trials, A&AEE for radio, Orange Harvet and sonobuoy trials, RAE for towing and parachute drag trials, SOC 31/01/73, used for fire fighting and rescue training at Farnborough.

WR973   First flight 18/01/57. Woodford for manufacturer's Mk 10 autopilot trials, MofS Air Fleet, A&AEE, No.49 MU. Woodford for performance. ventilation and vibration trials (cancelled), Armstrong-Whitworth for fuel system trials, Woodford (fuel system equipment removed), Phase 1 mods, No.203 Sqn, No.49 MU (Phase 2 mods), No.23 MU, No.203 Sqn, No.206 Sqn, Phase 3 mods and Viper installation, MofA for Viper flight trials, A&AEE for performance trials, Woodford. A&AEE service and hot weather trials (in USA), Woodford, Langar (for removal of trial equipment), MOD(Air), No.206 Sqn, Kinloss Wing, Bitteswell for mods, No.42 Sqn, used for fire fighting training at RAF Thorney Island, SOC as scrap June 1971.

WR974   First flight 1/05/57. C(A) on loan, A&AEE for tropical and winter trials, No.49 MU for winter trials equipment, A&AEE, CEPE Canada for winter trials, A&AEE, No.49 MU(Phase 1 mods), No.23 MU, No.203 Sqn, Langar (Phase 2 mods), MofA Air Fleet, Phase 3 trial installation, A&AEE for navigation and radio trials, Langar, A&AEE for tropical trials, Phase 3 equipment installation and Griffon 58 oil cooler mods, Langar, A&AEE for stand-by bomb bay heating, sonobuoy homer and photographic flash unit acceptance trials, ASWDU, Bitteswell for mods, Ministry of Technology (loan) for sonobuoy trials, RAE for radio trials, A&AEE for armament trials, ASWDU, Phase 3 mods, No.203 Sqn, No.42 Sqn (on loan), No.203 Sqn, Bitteswell for mods, No.203 Sqn, Kinloss Wing, No.2 SofTT as Instructional Airframe No.8117M, sold to the Peter Vallance Collection in 1988, extant.

WR975   First flight 26/05/57. RAF Handling Squadron, emergency landing at Exeter due to fuel leak, Woodford, RAF Handling Squadron, No.23 MU, No.220 Sqn, No.201 Sqn, No.49 MU (Phase 1 mods), No.203 Sqn, No.49 MU (Phase 2 mods), No.203 Sqn, No.201 Sqn, Phase 3 mods, No.206 Sqn, No.120 Sqn, Langar for Viper installation, Kinloss Wing, Bitteswell for mods, Kinloss Wing, No.32 MU, SOC and scrapped 1/10/71.

WR976   First flight 19/07/57. No.23 MU, No.220 Sqn, No.201 Sqn, No.49 MU (Phase 1 mods), No.206 Sqn, Phase 2 mods, No.206 Sqn, Langar for propeller strain-gauge trials, Phase 3 mods, No.201 Sqn, Langar for Viper installation, No.201 Sqn, Kinloss Wing, crashed into sea off Lands End 19/11/67, SOC as missing 19/11/67.

WR977   First flight 31/08/57. No.23 MU, No.220 Sqn, No.201 Sqn, Phase 1 mods, No.206 Sqn, Phase 2 mods, No.201 Sqn, Langar for Phase 3 mods and Viper installation, No.42 Sqn, Bitteswell for mods, No.2093 Sqn, No.206 Sqn (on loan), No.42 Sqn (on loan), No.203 Sqn, RAF Thorney Island for fire fighting training (cancelled), RAF Finningley museum as Instructional Airframe No.8186M, to Newark Air Museum 1/5/77, extant.

WR978   First flight 9/57. No.23 MU, No.220 Sqn, No.201 Sqn, No.49 MU (Phase 1 mods), No.206 Sqn, Phase 2 mods, No.206 Sqn, Phase 3 mods, No.42 Sqn, Woodford for Viper installation, No.42 Sqn, Kinloss Wing (on loan), RAF Fire Fighting School Catterick, SOC as scrap 1971.

WR979   First flight 1/11/57. No.23 MU, No.220 Sqn, No.201 Sqn, Phase 1 mods, No.201 Sqn, Langar (Cat.4 repairs and Phase 2 mods), No.206 Sqn, Phase 3 mods, No.120 Sqn, Kinloss Wing, Langar for Viper installation, Kinloss Wing, No.32 MU, SOC as scrap and broken up at St. Athan 1/10/71.

WR980   First flight 13/11/57. No.23 MU, No.220 Sqn, No.201 Sqn, Phase 1 mods, No.206 Sqn, Phase 2 mods, No.206 Sqn, No.120 Sqn, Langar for Phase 2 mods and Viper installation, No.206 Sqn, Kinloss Wing, Bitteswell for mods, Kinloss Wing, No.5 MU, RAF Fire Fighting School Catterick, SOC as scrap 1971.

WR981   First flight 12/57. No.23 MU, No.206 Sqn, No.49 MU (Phase 1 mods), No.120 Sqn, No.49 MU (Phase 2 mods), No.203 Sqn, No.23 MU (repairs), No.203 Sqn, No.23 MU (repairs), No.203 Sqn, No.201 Sqn, Phase 3 mods, No.120 Sqn, Langar for Viper installation, Kinloss Wing, Bitteswell for mods, No.32 MU, to RAF Topcliffe as Instructional Airframe No.8120M (17/12/70) and subsequently scrapped.

WR982   First flight 2/58. No.23 MU, No.206 Sqn, MofA, Woodford for engine fading research, No.206 Sqn, No.49 MU (Phase 1 mods), No.203 Sqn, Phase 2 & 3 trial mods, MofA Air Fleet, A&AEE for C(A) Phase 3 armament auxiliary fuel tank jettison and vibration trials, Langar for refurbishment, No.120 Sqn, Langar for Viper installation, No.201 Sqn, Kinloss Wing, No.60 MU (radio altimeter installed), Ministry or Technology, A&AEE for Viper water-methanol take-off and radio altimeter clearance trials, Kinloss Wing, withdrawn from service, No.2 SofTT Cosford as Instructional Airframe No.8106M, sold to Mr N. Martin Lutterworth in 1988 and subsequently scrapped.

WR983   First flight 3/58. No.23 MU, No.206 Sqn, No.49 MU (Phase 1 mods), No.206 Sqn, Phase 2 mods, No.206 Sqn, Phase 3 mods, No.120 Sqn, Kinloss Wing, Langar for Viper installation, Bitteswell for mods, Kinloss Wing, No.5 MU, broken up in 1970 with parts going to No.39 MU before final disposal.

WR984   First flight 6/03/58. No.23 MU, No.206 Sqn, No.49 MU (Phase 1 mods), No.203 Sqn, No.49 MU (Phase 2 mods), No.203 Sqn, No.201 Sqn, Phase 3 mods, No.120 Sqn, Langar for Viper installation, No.120 Sqn, Kinloss Wing, No.42 Sqn, Bitteswell for mods, RAF Topcliffe as Instructional Airframe No.8115M, SOC July 1971.

WR985   First flight 4/58. No.23 MU, No.206 Sqn, A&AEE for Maritime Tactical Position Indicator clearance trials, No.206 Sqn, No.49 MU, Phase 1 mods, No.203 Sqn, No.49 MU (Phase 2 mods), No.206 Sqn, Phase 3 mods, No.206 Sqn, No.60 MU (repairs), No.206 Sqn, No.120 Sqn, Langar for Viper installation, No.201 Sqn, Kinloss Wing, No.2 SofTT Cosford as Instructional Airframe No.8103M, sold to Jet Aviation Preservation Group in 1988, extant.

WR986   First flight 4/58. No.23 MU, No.206 Sqn, Phase 1 mods, No.203 Sqn, No.49 MU (mods and repairs), No.201 Sqn, Phase 2 mods, No.120 Sqn, Phase 3 mods, No.203 Sqn, Bitteswell for mods, scrapped due to rat infestation 1/9/71 and broken up at No.132 MU.

WR987   First flight 5/58. No.23 MU, No.220 Sqn, No.201 Sqn, No.120 Sqn, No.49 MU (Phase 1 mods), No.120 Sqn, Phase 2 mods, No.120 Sqn, Phase 3 mods and Viper installation, No.203 Sqn, Bitteswell for mods, No.5 MU, used for fire practice at RAF Honington, SOC as scrap June 1972.

WR988   First flight 5/58. No.23 MU, No.120 Sqn, Phase 1 mods, No.203 Sqn, Phase 2 mods, No.201 Sqn, Phase 3 mods and Viper installation, No.203 Sqn, No.42 Sqn (loan), No.5 MU, used for fire practice at RAF Macrihanish and SOC 25/04/72.

WR989   First flight 6/58. No.23 MU, No.120 Sqn, No.49 MU (Phase 1 mods), No.120 Sqn, Phase 2 mods, No.201 Sqn, Phase 3 mods, No.201 Sqn, No.120 Sqn, Langar for Viper installation, Kinloss Wing, No.60 MU for radar altimeter installation, Ministry of Technology (loan), A&AEE for Viper high-humidity and low-temperature flight trials, Kinloss Wing, Bitteswell for mods, No.203 Sqn, Bitteswell for mods, No.203 Sqn, No.5 MU, used for fire fighting practice at RAF Leeming and SOC 14/07/72.

WR990   First flight 7/58. No.23 MU, No.120 Sqn, No.49 MU (Phase 1 mods), No.120 Sqn, No.49 MU (Phase 2 mods), No.120 Sqn, Phase 3 mods, No.201 Sqn, Langar for Viper installation, Kinloss Wing, No.60 MU for radar altimeter installation, Kinloss Wing, RAF Newton as Instructional Airframe No.8107M and subsequently SOC 17/10/70.

Specification Number:
Air Ministry 42/46
Contract Number:
6/ACFT.6408/CB6(a), contract amended in September 1953
Shackleton MR Mk 3
Additional quantity: of seventeen, only thirteen built (XF700 to XF711 & XF730 (XF731 to XF734 were cancelled before construction started on them).

XF700   First flight 9 /7/58. No.23 MU, No.120 Sqn, No.49 MU (repairs and Phase 1 mods), No.120 Sqn, Phase 2 mods, No.120 Sqn, Phase 3 mods, No.206 Sqn, Langar for Viper installation, Kinloss Wing, Bitteswell for mods, No.203 Sqn, used for fire fighting training in Nicosia and SOC as scrap 26/10/71. Some parts ended up at No.103 MU but these too were scrapped in January 1972.

XF701   First flight 8/58. No.23 MU, No.120 Sqn (repairs and Phase 1 mods), MofA (loan), A&AEE for acceptance trials, Phase 2 mods, A&AEE for mod retrial, No.206 Sqn, Phase 3 mods and Viper installation, No.206 Sqn, Kinloss Wing, A&AEE, Bitteswell for mods, No.42 Sqn, Bitteswell for mods, CTE Manston and SOC as scrap 13/08/71.

XF702 - First flight 9/58. No.23 MU, No.203 Sqn, No.49 MU (Phase 1 mods), No.120 Sqn, No.49 MU (Phase 2 mods), No.206 Sqn, Phase 3 mods, MofA (loan), RAE for Safety Study Group review, A&AEE for GM7 compass trials, No.206 Sqn, No.120 Sqn, Viper installation, Kinloss Wing, crashed at Creag Bhan Inverness 21/12/67 and SOC.

XF703   First flight 9/58. No.23 MU, No.203 Sqn, No.49 MU (Phase 1 mods), No.120 Sqn, No.203 Sqn, Phase 2 mods, No.206 Sqn, Phase 3 mods, No.206 Sqn, HSA(Loan), No.201 Sqn, Kinloss Wing, Viper installation, Kinloss Wing, A&AEE for ATN-71 and STR-40A radio altimeter installation and flight tests, Kinloss Wing, No.42 Sqn, Bitteswell for mods, No.42 Sqn, RAF Henlow for the RAF Museum, subsequently scrapped there.

XF704   First flight 10/58. No.203 Sqn, No.23 MU, No.203 Sqn, No.49 MU (Phase 1 mods), No.120 Sqn, No.49 MU (Phase 2 mods), No.203 Sqn, No.201 Sqn, Phase 3 mods, No.206 Sqn, Kinloss Wing, Bitteswell for installation of Ferranti stall-warning system, Woodford for Ferranti anti-stall warning flight trials, A&AEE for anti-stall warning clearance trials, Kinloss Wing, Bitteswell for mods, withdrawn from service, CTE Manston and SOC August 1971.

XF705   No information available.

XF706   First flight 12/58. No.203 Sqn, No.49 MU (Phase 1 mods), No.120 Sqn, No.49 MU (Phase 2 mods), No.203 Sqn, No.23 MU, No.201 Sqn, Phase 3 mods, No.42 Sqn, Viper installation, Bitteswell for mods, Kinloss Wing (on loan), No.42 Sqn, withdrawn from service, allocated Instructional Airframe No.8089M but never taken up, used for fire fighting

practice at RAF St. Mawgan and burnt in March 1970.

XF707   First flight 1/59. A&AEE for ASV 21 radar cooling trials, Phase 1 mods, No.201 Sqn, Phase 2 mods and repairs, No.206 Sqn, Phase 3 mods, No.42 Sqn, Viper installation, No.42 Sqn, withdrawn from service, used for fire fighting practice at RAF Benson and SOC as scrap 28/04/71.

XF708   First flight 1/59. Phase 1 mods, No.201 Sqn, Phase 2 mods, No.120 Sqn, Phase 3 mods, No.203 Sqn, No.5 MU, to the Imperial War Museum Duxford 23/08/72, extant.

XF709   First flight 3/59. Phase 1 mods, No.201 Sqn, Phase 2 mods, No.120 Sqn, Phase 3 mods and Viper installation, No.203 Sqn, HSA for new tailplane de-icing system trials, Kinloss Wing, Bitteswell for mods, No.32 MU, NEA, SOC as scrap 1/10/71.

XF710   First flight 3/59. Phase 1 mods, No.201 Sqn, Phase 2 mods, No.120 Sqn, crash-landed on Culloden Moor Inverness 10/01/64 and SOC as scrap.

XF711   First flight 4/59. Phase 1 mods, No.201 Sqn, SBAC display doing 22-hour sorties each day, Phase 2 mods, MofA Air Fleet (on loan) for new Phase 3 trials, Woodford for rebuild, A&AEE for tropical trials at Idris, RAF St. Mawgan for take-performance trials, A&AEE for handling techniques, Phase 3 mods, No.42 Sqn, Viper installation, Kinloss Wing, No.42 Sqn, Bitteswell for mods, No.42 Sqn, withdrawn from service, used for fire fighting practice at RAF Abingdon and SOC as scrap 7/06/71.

XF730   First flight 5/59. Phase 1 mods, No.206 Sqn, No.201 Sqn, Phase 2 mods & repairs, No.120 Sqn, Phase 3 mods, and Viper installation, A&AEE for trials, withdrawn from service, used for fire fighting practice at RAF Kinloss and SOC as scrap June 1971.

Specification Number:
Air Ministry 42/46
Contract Number:
6/ACFT.11106/CB6(a), dated July 1953
Shackleton MR Mk 3
Quantity: 13, Serial numbers XG912 to XG924 was cancelled in February 1956.

Specification Number:
Air Ministry 42/46
Contract Number: B1/8126, dated March 1954
Shackleton MR Mk 3
Quantity: 8 for South African Air Force
S/Nos 1716 to 1723 (construction nos 1526 to 1533)

1716   First flight 29/03/57. Accepted by No.35 Sqn (SAAF), RAF St. Mawgan for SAAF work-up, departed for Cape Town (arrived 18/08/57), wing re-sparred between March 1973 and April 1976, SAAF Museum Swartkop, made airworthy in 1994 but crashed in Sahara Desert en route to the UK on the 13th July 1994.

1717   First flight 6/05/57. Accepted by No.35 Sqn (SAAF), RAF St. Mawgan for SAAF work-up, departed for Cape Town (arrived 18/08/57), wing re-sparred between September 1975 and October 1977, withdrawn from service, Ysterplatt AFB (stored), to Natal Parks Board Museum, sold to businessman from Stranger, Kwazulu, Natal. Extant.

1718   First flight 13/05/57. Accepted by No.35 Sqn (SAAF), RAF St. Mawgan for SAAF work-up, departed for Cape Town (arrived 18/08/57), wheels-up landing at D.F. Malan Airport, repaired, crashed in Wemmershook Mountains and SOC 8/08/63.

1719   First flight 6/09/57. Accepted by No.35 Sqn (SAAF), RAF St. Mawgan for SAAF work-up, departed for Cape Town (arrived 13/02/58), returned to UK for training with Coastal Command, back to South Africa, withdrawn from service (4/78), Ysterplaat AFB (stored), displayed outside Stellenbosch airfield clubhouse, then displayed on waterfront complex in Cape Town (1991) but subsequently scrapped.

1720   First flight 26/09/57. Accepted by No.35 Sqn (SAAF), RAF St. Mawgan for SAAF work-up, departed for Cape Town (arrived 13/03/58), damaged at D.F. Malan Airfield (9/61), repaired, with drawn from service (3/83), displayed at Ysterplatt AFB outside Officer's Club.

1721   First flight 12/12/57. Accepted by No.35 Sqn (SAAF), RAF St. Eval, departed for Cape Town (arrived 26/02/58), wheels-up landing at Ysterplaat AFB (9/62), repaired, took part in retirement flight at D.F. Malan Airport 23/11/84, displayed at SAAF Museum, Swartkop, extant.

1722   First flight 7/02/58. Accepted by No.35 Sqn (SAAF), departed for Cape Town (arrived 26/02/58), nose wheel collapsed at Langebaanweg (6/60), repaired, returned to UK for JASS course in June 1964, back to South Africa (30/07/64), took part in retirement flight at D.F. Malan Airport 23/11/84, aircraft currently retained by No.35 Sqn in flying condition for the SAAF Museum (due to be grounded in 2005?)

1723   First flight 10/02/58. Accepted by No.35 Sqn (SAAF), departed for Cape Town 14/02/58, hydraulic failure en route and diverted to Ysterplaat AFB running off runway on landing (26/02/58), repaired, withdrawn from service (22/11/77), stored at Ysterplaat AFB, sold to Mr Vic de Villiers and now mounted on the roof of his garage ('Vic's Viking'

garage) on the highway to Johannesburg painted like a giant Coca-Cola bottle! Extant.

Note: Serial numbers **emboldened** are still extant as at February 2005.

A formation of three Shackletons off the North coast of England. It is interesting to see that only the lead aircraft is still fitted with the forward twin cannon.
via Martyn Chorlton

Appendix IV: SHACKLETONS TESTED BY THE A&AEE BOSCOMBE DOWN

Quite a number of Shackletons were tested at Boscombe Down over the years, so here is a list of those involved and what trials work they undertook.

MR Mk 1

• VW126

This aircraft visited in 1950 and undertook Carbon-Dioxide contamination tests as well as assessment of the new spring tabs on the rudder that replaced the previous geared system. The noise levels in the cockpit were also reduced by the fitment of a ducted exhaust system to replace the original stub exhaust stacks.

• VW131

This aircraft was first at A&AEE in February 1950 and undertook all the initial prototype trials for the type. It was found to be a good aircraft in most conditions, although it was tiring for the crew due to heavy controls in turbulent conditions and difficult to taxi, especially in a crosswind. The noise from the inboard contra-rotating propeller was considered to be high for the pilot and the rudder was heavy in its operation and overbalanced at high angles of attack. The lack of an accurate stall warning system was highlighted and of course this was to have dire consequences with the MR Mk 3 many years later. Performance data from this aircraft while at A&AEE was as follows.

Minimum speed on port engines alone with 9psi boost - 105 knots.

Take-off distance at 86,000lb with 17° of flap - 910 yards (unstick), 1,240 yards (to clear 50ft).

Rate of Climb - 940 ft/min.

Minimum landing run at 72,000lb - 1,100 yards or 1,260yards for an instrument approach.

Hot Weather Trials (Khartoum).

Take-off distance at 86,000lb 'dry' - 1,635 yards.

Take-off distance at 86,000lb 'wet' (with water-methanol injection) - 1,180 yards.

Rate of Climb - 655 ft/min.

The aircraft was completely re-rigged in mid-1951 during tests to try and reduce vibrations and variations during flight.

• VW135

This aircraft undertook flights of up to 15½ hours and in the report written by the A&AEE staff they concluded that the type was 'unacceptable for long range maritime operations' due to the tendency of the aircraft to yaw and the levels of vibration and noise. It was also involved in armament trials and clearance trials for all the bomb bay ordnance combinations from 11lb right up to 1,000lb. It also performed radio and radar trials and cleared all such equipment intended for the type at this time, including ASV Mk 13, during 1950-1.

• VP254

This aircraft was involved in electrical equipment trials and was used to clear for service use IFF 10 and SARAH rescue equipment, once defects with the latter had been ironed out.

• VP255

This aircraft was involved in armament trials and it was in this machine that it was discovered that the gunner could not get into the dorsal B.17 turret while wearing full flying clothing. The distance between airframe and expended rounds from the upper turret was also found to be too small, thus increasing the risk of inadvertent damage to the aircraft when the guns were fired. In 1956 this aircraft also carried out torpedo, depth charge and sonobuoy clearance for the type. Finally in 1956 Lindholme rescue equipment was also cleared thanks to this aircraft.

• VP259

This aircraft undertook operational assessment and was seen to weigh 85,739lbs with 3,292 Imp. Gal. of fuel and 5,500lb of armament. This gave a calculated range of 2,670 miles, thus meeting the Air Ministry specification for the type.

• VP261

This aircraft had the elevator span increased by 20% and the balance tab ratio changed during early 1951, all in an attempt to improve flight characteristics.

• VP284

This aircraft was involved in electrical equipment trials during mid-1955 and it tested the Blue Silk (doppler) and the GPI 4a (later GPI 6) ground position calculator systems prior to service adoption.

• VP293

Long flights undertaken in this aircraft in 1952 echoed the comments made in the assessment of VM135 in 1950, and the inadequate toilet facilities were extremely taxing for the 13-man crew on such flights! This aircraft also undertook flight accuracy checks and it was found that the type wandered quite considerably in flight and that the Mk 9 autopilot made little difference. The vibration and noise levels were all above accepted levels and the former resulted in making it almost impossible for the navigator to work on his plotting table, as it vibrated so much!

• WB835

This aircraft was involved in testing the heating system in late 1952, which was found to be insufficient as a result.

MR Mk 2

• VW126

The prototype Mk 1 was soon back as the aerodynamic prototype for the Mk 2. In May 1952 the type was tested at A&AEE and the new lockable tailwheel and air brakes were all greeted well by the test crews, albeit that the air supply for the latter was soon used up and there was no way of replenishing it. Once again the type still suffered from unacceptable noise and vibration levels. The stall warning was still insufficient and directional stability was not improved in relation to the Mk 1.

• WB833

This aircraft was tested with the CofG further aft, although this made control forces near the stall to be reversed, so the new layout was not adopted and the old configuration remained. In 1958 this aircraft was used to test ASV 21 as well as LORAN hyperbolic navigation aid. This was followed in 1960 by tests of UHF radio and TACAN navigational aids.

• WG530

This aircraft adopted the improved oil cooling system during trials in Khartoum. In 1953 it was fitted with leading edge spoilers, underwent stall trials and although under power this system gave a better warning of the onset of the stall, at cruise there was no warning at all. This machine also did most of the armament trials for the Mk 2. These included testing sonobuoy chutes (Mk 9 and 11) as well as the new Boulton Paul Type N nose turret, which was found to have unacceptable levels of vibration effecting the sighting system when the guns were fired.

• WG532
This aircraft undertook investigations into dive recovery limits and a maximum load of 1.8g was achieved with the type during pull-out.

• WG558
This aircraft was used to test a new system of flare chute during 1961.

• WL759
This aircraft did test drops with a Mk 44 torpedo in 1963-4.

• WL785
This aircraft was used to evaluate the use of parachutes on sonobuoys.

• WR953
This machine was involved in electronic tests and it was discovered that IFF 10 had a poor range during trials in this aircraft. The new radio compass was also tested in this aircraft and it was discovered that the location of the associated aerials was a critical consideration.

• WR962
This aircraft undertook trials with the test bomb bay load comprising three Mk 30 torpedoes and an oxygen crate.

• WR964
This aircraft undertook yet more cooling trials during mid-1954, although this time in Aden.

• WR965
This aircraft was used for a revision to the armament system in 1962.

MR Mk 3

• WR970
The prototype was tested at Boscombe Down by a team from Avro and using members of B Squadron at the A&AEE, they were able to cut down on the service trials needed on the type. This aircraft was with the A&AEE from September 1956, but it ultimately crashed fatally in November after a modified stall indicator had been fitted at Woodford.

• WR971 & WR972
Both of these machines arrived with the A&AEE in December 1956. They undertook initial Controller (Air) tests and then went on to work on performance and armament trials that went on until mid-1958.

• WR973
In 1965 this machine was test fitted with the Viper jet engines in the outer engine nacelles and a team from the A&AEE visited Avro at Woodford to assess this installation.

• WR974
Once this aircraft had been updated to Phase 3 standard it was used for trials in the 1964-5 period.

• WR982
This machine was used for a considerable period to test the mass of potential weapon bay configurations and to do air drops on all ordnance. The initial period lasted until April 1964, but in 1967 the aircraft came back to the A&AEE to be used to investigate the effects on performance that would be encountered should the Viper jet engines be allowed to stop and windmill in the airflow.

• WR985
It is believed that this machine was used to test an improved aerial reflector box, although precise records are no longer available.

• XF701
In 1962 this machine tested the new AD712 radio compass installation.

• XF702
Once this aircraft had been updated to Phase 3 standard it was used for trials in the 1964-5 period.

• XF703
This machine was used for a short series of trials in 1968 to test a radio altimeter made by Honeywell.

• XF705
This machine was used for a series of trials relating to the improved stall warning system during 1969.

• XF711
This aircraft was tested and showed an increase in all-up weight to 106,000lb.

AEW Mk 2

• WL745
The prototype was used to do initial service trials and test. These started in early 1971 and the all-up weight on this type had risen to 96,000lb and a maximum speed of 300 knots was imposed in consideration of the strength of the new ventral radome.

• WL757
This machine was used to test improved UHF aerial positions during 1974.

• WL793
This aircraft made a brief visit to the A&AEE during August 1973 for checks on crew emergency access routes.

T Mk 4

• VP258
This aircraft, converted from a MR Mk 1, was used to do the initial type tests in early 1957.

• VP293
This aircraft continued type trials with WB858 during 1959. It carried operational equipment such as armament and avionics and was more like the operational versions they were meant to help train crews with.

• WB858
This aircraft continued type trials with VP293 during 1959. This aircraft also had the avionic and armament equipment seen on VP293.

THE AVRO SHACKLETON 107

PELICAN 16 – LAST FLIGHT OF THE SHACKLETON

Once the Shackleton had retired from squadron service in the UK it seemed as if the type had disappeared from our skies forever. However, in 1993 plans were put in place by a number of organisations to bring a SAAF MR Mk 3 across to the UK. Sponsorship had been arranged for fuel etc and FLS Aerospace (Lovaux) Ltd at Bournemouth in the UK had agreed to offer technical support. By mid-1994 these plans had been finalised and Shackleton S/No. 1716 was allocated the call-sign Pelican 16 for the flight. This call-sign is a combination of the No.35 Squadron badge, which is a Pelican, and the last two numbers of the aircraft serial number. The aircraft, captained by

Major Eric 'Pine' Pienaar, was prepared for the flight and left D.F. Malan airfield with nineteen people on board. The journey would go on a set route via Liberville (Gabon), to Abidjan (Ivory Coast) then Lisbon and finally to Duxford for its first display. The aircraft took-off on the 7th July 1994, but this was already a day behind schedule and the aircraft developed a problem with the No.1 (port outer) engine during the leg from Liberville to Abidjan. The problem happened during the final stages of this flight, so the engine was shut down and the aircraft made a safe landing at Abidjan on three. Engineer Gus Guse inspected the engine and stated that a replacement was advisable so one of the two spare engines held back at the SAAF Museum collection at Swartkop was flown out. It was

installed and ground run and the aircraft took off for the long leg to RAF Lyneham via Lisbon at 7pm on the 12th July. This was a long 'dog-leg' route via Las Palmas to avoid the Atlas Mountains and at 1am on the 13th July while the aircraft was flying over the Sahara Desert trouble developed. Initially the number four (starboard outer) engine overheated and had to be shut down. The pilot set a new course to try and use the cooler air over the Atlantic, but thirty minutes later number three engine started to develop problems in the gearbox and was also shut down. The propeller on number three continued to windmill and although number 4 was restarted, it soon started to overheat again. The air temperature that night was extremely high and it was also very dark, which is why the decision had been made to try and get to the cooler air over the ocean. Now the aircraft had both the number three and four engines on the starboard side shut down and with both sets of propellers windmilling, the aircraft was losing height rapidly. The huge asymmetric effect of the two contra-rotating propellers meant that the pilot was struggling all the time to keep the aircraft on a straight heading. With the height now down to 4,000ft the decision was made to make an emergency landing and at 1.30am a Mayday signal went out, stating 'Mayday, Mayday, Mayday, Pelican 16 heading 305 degrees magnetic. Crash landing, Two engines out'. This call was intercepted by an Air France airliner and passed on. All SAAF Shackletons got hydraulic pressure via the pumps on the Nos.3 and 4 engines, but the windmilling propellers on these engines produced enough pressure for the flaps to be lowered to the take-off position. The aircraft was now descending at a rate of 400ft per minute and the crash landing was achieved at 1.35am (GMT) with nothing less than the navigator, his topographical map and radio altimeter and the co-pilot who acted as look out for small hills etc during the run-in. The tail touched down first (although some sources state it was the radar scanner that touched down first), then the entire aircraft slid for 200 metres. Engine number four was torn out, the undercarriage was ripped off when it struck the only sizable rock in the whole area(!) and the fuselage was badly buckled. Despite all this the crew was uninjured and they got out of the aircraft quickly because the fuel tanks had been ruptured and aviation fuel was spilling onto the ground. The crew set up the emergency radio beacon and recovered the food and blankets etc from the aircraft before making camp.

The secretary of the South African Embassy, Mr Onnie Kok was woken at 6am to be told that the aircraft had crashed and in co-operation with the French

This shot shows SAAF MR Mk 3 S/No.1722 starting up in 2004.

A shot of airworthy MR Mk 3 S/No.1722 taxying out prior to a flight during 2004.
Both Andrew White

Air Force and Moroccan civil aviation authorities they were able to locate the aircraft and its stranded crew by 7.30am. The crew of the downed aircraft had doused the ripped off undercarriage with hydraulic

A still from video footage taken by the crew on the 13th July 1994.

Taken in January 2005 by NATO forces, this photograph shows 1716 as she is today, slowly being claimed by the sands.

View of the inboard port engine, with the blades bent form the impact and the lower fuselage skin buckled.
All SAAF Museum

fluid and set it on fire to act as a signal, and the first aircraft to locate them was a Breguet Atlantique of the Maritime Nationale. This aircraft dropped a canister of medical supplies and circled them for about 30 minutes before moving off to co-ordinate the UN border patrol towards the crash site. The Shackleton had come down in a disputed border area between Mauritania and the Western Sahara, about 50km from the border town of Zouerat. As a result the crew were airlifted by a United Nations peacekeeping force Bell helicopter to the Hasi Augenit oasis. The tense political situation in this region meant that the crew remained at the oasis while their return to Cape Town was sought. In this time the crew received the unusual gift of a goat from the Polisario Front guerrillas, who having heard of their survival reckoned that they had been given a second chance by the gods, and were therefore happy to help them! The crew were on their way back home the next day and they staged through Tindouf in Algiers before eventually getting back to Cape Town in an SAAF Boeing 707 on the 17th July. Not all of the crew went though, as the SAAF Public Relations Officer, Col. Derrick Page, made his way to the UK, and kept his promise by being at the Duxford airshow where 1716 should have made her first public appearance.

It was a sad end to what many thought was to be the last flight of a Shackleton and to this day Pelican 16 still lies in the desert where she came to rest back in 1994; 2838N, 1314W. Although there are still a number of Shackletons around the world (see list on page 61), only two are airworthy. The SAAF Museum at Ysterplaat AFB in Cape Town, South Africa has S/No.1722, which is kept in airworthy condition. This aircraft was flown three times on the 13th July 2004 to mark the 10th anniversary of the loss of Pelican 16. It is an aircraft that is behind on flying hours per year, so the three flights undertaken that day were to keep the three pilots current on the type. This aircraft is currently under threat of grounding, due to lack of spares and money, so with the hammer apparently having been due to fall during 2004 we will just have to wait and see what develops for this aircraft during 2005. The only other Shackleton airworthy in the world is AEW Mk 2, WL790 (N790WL) which is called 'Mr McHenry' and is owned by Air Atlantique. Due to UK aviation authority regulations the type cannot be operated or based in the UK, so this aircraft is currently operated by Amjet at Blaine, Minnesota. Air Atlantique do state on their website that they intend to return this aircraft to the UK, although as I write there was no indication as to when this might happen. Although it is a complex and expensive aircraft to operate, it is a shame that such a historic type, which served for so long and so well, does not even fly in the country of its birth. At least if you are in the USA or South Africa you can still see a Shackleton up there in the sky, where she belongs and experience for yourself the true nature of 'the Growler'!

GLOSSARY

A&AEE	Aeroplane & Armament Experimental Establishment	knot	Unit of speed of 1 nautical mile per hour (approx 1.15mph or 1.85km/h)
ACM	Air-Chief Marshal (RAF)	LAC	Leading Aircraftsman
AC1	Aircraftsman 1 (RAF)	lb	Pound
AC2	Aircraftsman 2 (RAF)	lt	Litre
AEW	Airborne Early Warning	MAD	Magnetic Anomaly Detection
AFB	Air Force Base	Maj	Major
Air Cdre	Air Commodore (RAF)	MAP	Ministry of Aircraft Production
Air Mshl	Air Marshal (RAF)	MEAF	Middle East Air Force
AOC	Air Officer Commanding	Mk	Mark
A.S.	Anti-Submarine	MofA	Ministry of Aviation
ASR	Air-Sea Rescue	MOTU	Maritime Operational Training Unit
ASV	Air to Surface Vessel (Radar)	mph	Miles Per Hour
ASWDU	Air-Sea Warfare Development Unit	MU	Maintenance Unit (RAF)
AVM	Air Vice Marshal (RAF)	NATO	North Atlantic Treaty Organisation
AWACS	Airborne Warning And Control System	No.	Number
BS	British Standard	NCO	Non-commissioned Officer
Capt	Captain	OC	Officer Commanding
CENTO	Central Treaty Organisation	OCU	Operational Conversion Unit
CFS	Central Flying School	OTU	Operation Training Unit
CGS	Central Gunnery School	Plt Off	Pilot Officer (RAF)
CinC	Commander-in-Chief	RAE	Royal Aircraft Establishment
CO	Commanding Officer	RAF	Royal Air Force
CRD	Civilian Repair Depot	RAE	Royal Aircraft Establishment
CSDE	Central Servicing Development Establishment	RRE	Radar Research Establishment
		RSU	Repair & Servicing Unit
DF	Direction Finding	SAAF	South African Air Force
DFC	Distinguished Flying Cross	SAC	Senior Aircraftsman (RAF)
DFM	Distinguished Flying Medal	SAR	Search & Rescue
DSO	Distinguished Service Order	SARAH	Search And Rescue Automatic Homing
DTD	Directorate of Technical Development	SBAC	Society of British Aircraft Construction
ECM	Electronic Counter Measures	Sgt	Sergeant
FAA	Fleet Air Arm	SOC	Struck Off Charge
FEAF	Far East Air Force	SoTT	School of Technical Training
Fg Off	Flying Officer (RAF)	Sqn	Squadron
Flt	Flight	Sqn Ldr	Squadron Leader (RAF)
Flt Sgt	Flight Sergeant	T	Trainer
Flt Lt	Flight Lieutenant	TACAN	Tactical Air Navigation
FS	Federal Standard	TRE	Telecommunications Research Establishment
ft	Foot	UK	United Kingdom
g	Acceleration of free fall due to gravity	USAF	United States Air Force
GHQ	General Headquarters	USS	United States Ship
Gp Capt	Group Captain (RAF)	VC	Victoria Cross
H.E.	High Explosive	Wg Cdr	Wing Commander (RAF)
HMS	His/Her Majesty's Ship	W/O	Warrant Officer
HQ	Headquarters	W/T	Wireless Telegraphy
IAM	Institute of Aviation Medicine	WWI	World War I (1914-1918)
IFF	Identification Friend or Foe	WWII	World War II (1939-1945)
in.	Inch		
JASS	Joint Anti-Submarine School		
kg	Kilogram		
km	Kilometre		
km/h	Kilometres Per Hour		

INDEX

FINAL RESTING PLACE

This photograph taken many years ago at Kinloss shows the sad remains of a MR Mk 1.

Stripped of all useable parts, this unidentified MR Mk 1 is seen on the dump at Kinloss many years ago.

This MR Mk 3 is seen here on the dump at Kinloss stripped of all useable parts.
All John Woodward

MR Mk 3, XF701 makes a sad sight at the CFE Manston where it was eventually SOC as scrap in August 1971. The lack of cowlings on the outer engine nacelle in this shot shows the manner in which the Viper engine was suspended below and behind the Griffon.
via Martyn Chorlton